Builder of the Past

—

Architect of the Future

THE HISTORY OF THE REA/RUS TELEPHONE PROGRAM

Linda M. Buckley
Research and Writing

Complete Graphic Services
Design and Production

Foundation
for Rural Service

This book was published by the Foundation for Rural Service, a non-profit organization established by the National Telephone Cooperative Association to inform and educate the public on the rural telecommunications industry and to improve the quality of life throughout rural America. Publication of the book was made possible by the generous support of the Rural Telephone Finance Cooperative.

Printed in the United States of America

ISBN: 0-9674251-0-7

This book is dedicated to the men and

women of independent telephony,

the REA/RUS staff, and the policy

makers—both past and present—

whose innovation, perseverance, and

commitment have made this nation's rural

telecommunications system

the best in the world.

S everal years ago, the leaders of the Foundation for Rural Service (FRS) identified the upcoming 50-year anniversary of the Rural Electrification Administration (REA)/Rural Utilities Service (RUS) telephone loan program as an event that should not pass unnoticed. Realizing the significance of the loan program to National Telephone Cooperative Association (NTCA) members' beginning, growth, and continued success—as well as to the association's own formation—they set out to appropriately commemorate this historic milestone.

The FRS, established by the NTCA to inform and educate the public on the rural telecommunications industry and to improve the quality of life throughout rural America, was the obvious entity to spearhead such a commemorative project. The FRS board of directors determined a showcase book, telling the story of the REA/RUS program and its borrowers in both words and photographs, would best capture the spirit of the past 50 years. The Rural Telephone Finance Cooperative (RTFC) graciously offered to provide financial support for the extensive project.

As the individual selected to research and write the book, I called upon the knowledge and assistance of people throughout the nation as I gathered documents, photographs, and memories of the telephone loan program. Going to various libraries and the National Archives, as well as sorting through scores of items at the RUS, I found the abundance of information, particularly from the first 30 years, nothing short of overwhelming.

Also, countless people had a memorable REA/RUS story to tell. From telephone system owners and REA/RUS staff to industry experts and trade association staff, each person carefully recalled and generously shared his or her part in this marvelous history. Many provided materials —often one-of-a-kind, irreplaceable items—for me to use in completing the book. Several also suggested additional individuals with whom to speak, but, unfortunately, time did not permit me to speak to every one.

With the amount of material supplied and discovered, I quite easily could have written a multi-volume set. It proved quite a challenge to initially select which information and photographs to include in the book and an even greater challenge to whittle away details to make everything fit. But I worked with representatives of the FRS, NTCA, and RTFC, and together we met the challenge. I believe the end result is a fitting tribute to this important part of our nation's history.

I truly am honored to have had the privilege to research and write this book, to be entrusted with such valuable historic items, and to work with the many fine people who played a role in the book's production. I would like to take this opportunity to thank those individuals.

Rachel Brown provided editorial assistance for portions of the project, helping to conduct interviews, gather information, write borrower stories, and check facts.

My RUS contact, Legislative and Public Affairs Advisor Claiborn Crain, as well as his assistant Holly Howell, was of invaluable assistance, always ready to search for a document or answer for me, no matter how obscure my request.

Many REA/RUS staff members, both past and present, contributed information for the book. Several took time to talk with me or my assistant extensively, including Wally Beyer, Christopher McLean, Ed Cameron, Bob Peters, John Arnesen, Ira Shesser, John Rose, William Bird, Claude Buster, Lamar Moore, Richard Gabel, Cheryl Prejean Greaux, and Cheryl Gamboney. I also wish to thank those who provided information indirectly by answering the many questions I posed to Claiborn, as well as those who assisted in setting up appointments and locating items.

A special thank you to the REA alumni group and their leader William Morris for making me an honorary member during the project, inviting me to attend their monthly luncheons, and allowing me to speak about the project during two of their meetings. The stories they shared were wonderful, and I truly enjoyed my visits with the group.

Several industry representatives involved in REA/RUS issues through the years provided extensive information as well, including Shirley Bloomfield and Tom Wacker of NTCA; Bob Leigh, retired from NTCA; and John O'Neal of the National Rural Telecom Association.

The response from telephone systems in supplying documents, photographs, and information for the book was excellent. The number of contributors is too many to include here, but the names are listed at the end of the book.

I also would like to recognize FRS Director Lisa Lopinsky and members of the NTCA staff for their part in this book and thank them for being such a great group with whom to work.

Finally, I wish to thank my husband and daughter for their support throughout this project, as well as the many family members and friends who assisted me with child care and other miscellaneous items so I could spend additional time working on "the book."

I can honestly say this project is among the most interesting I've undertaken during my 15 years in the telecommunications industry. I thoroughly enjoyed learning so much about such a unique program and its benefits to rural Americans, and I truly admire the REA/RUS borrowers and staff for their dedication and accomplishments.

Linda M. Buckley
Researcher/Writer

The signature lines and text:

William H. Stanley Jr.

William H. Stanley Jr., *President*
Foundation for Rural Service

Sheldon C. Petersen

Sheldon C. Petersen, *CEO*
Rural Telephone Finance Cooperative

Michael E. Brunner

Michael E. Brunner, *Executive Vice President*
National Telephone Cooperative Association

*T*he dramatic advances we have seen during this century in the telecommunications industry have been nothing short of revolutionary. From a primitive local network and crank telephones to a world of wireless phones, cable television, and electronic chat rooms.

The history of this dynamic industry would not be complete without telling the story of how our nation's remote rural areas gained access to modern telecommunications technology. In an era of limited technology, America's rural residents banded together to form rural telephone systems that would bring the benefits of modern telephony to the farm, the village, and the country store. By doing what the large telephone companies were too profit-minded to consider, these community-spirited enterprises closed the loop on the nation's telecommunications network and ensured all Americans would have access to this essential utility service.

There are many heroes in this important tale, but chief among them is the federal government. Through the Rural Electrification Administration (REA)—now the Rural Utilities Service (RUS)—it supplied the capital, in the form of low-interest-rate loans, to make the dream of rural telephony a reality. Without this fundamental support, it's safe to say that rural America would not have as quickly developed the advanced telecommunications infrastructure that now is so vital to all aspects of everyday life.

In addition, substantial credit goes to those rural citizens who organized and built the local telephone systems. Their initiative helped ensure their friends and neighbors would not be left behind their urban counterparts as technology advanced and became more essential to ordinary business and social discourse. The accomplishments of these dedicated Americans cannot be underestimated or ever fully rewarded. We hope, however, that by committing to tell their story, we acknowledge their inestimable contribution to American life.

In the end, it is not where we have been, but where we are going that really matters. Yet the future always stands upon the shoulders of the past, and tomorrow's accomplishments are predicated upon how prior challenges have been met. As we move into the next century, let us take time to review our rich heritage and honor those individuals and institutions who have ensured the vast opportunities that lie ahead.

In 1949, only 39 percent of rural Americans were receiving even minimal telephone service. But that year, a new era for rural telephony began with the enactment of the legislation establishing the Rural Electrification Administration telephone loan program—now administered by the Rural Utilities Service (RUS) of the U.S. Department of Agriculture.

Today, 50 years later, the telephone loan program has developed into a successful, forward-focused program that meets a continuing, strong demand for investment capital. Thanks in large part to the work of RUS telecommunications borrowers, more than 94 percent of U.S. households now have telephone service, and nearly all subscribers of RUS borrowers are served by digital technology.

Through the deployment of advanced telecommunications technologies, RUS borrowers have produced a rural network of more than one million miles of telecommunications line, including thousands of miles of fiber optic cable, to serve millions of rural residents. By providing grants and loans for distance learning and telemedicine projects, the RUS also is providing rural residents information-age solutions to the problems facing rural educators and health care providers.

Advanced telecommunications services are crucial to rural America's economic development. Rural businesses must be able to compete, and rural residents need reliable access to the global economy, as well as quality health care and educational opportunities. The concept of "basic" voice telephone service no longer exists—the vision of universal service must include "core" services, defined in terms of speed and bandwidth, that evolve with new technologies. RUS borrowers are leading the way, as loans and grants made by the RUS program help advance universal service and affect the lives of millions of rural residents.

We will continue our public-private partnership to ensure the capital needs of the rural telecommunications industry are met. Together we will cross the bridge to the 21st century, making quality, reliable, and affordable telecommunications available to all rural citizens.

I salute all RUS telephone program participants for their dedication and innovation, and I extend my sincerest congratulations and appreciation to our country's rural telecommunications systems for these 50 years of quality service to rural Americans.

Wally Beyer, *Administrator*
Rural Utilities Service
U.S. Department of Agriculture

REA TELEPHONE MEETING

YOU ARE INVITED to a meeting sponsored by the MIDSTATE TELEPHONE COMPANY for the purpose of explaining plans for providing your area with a modern dial telephone system. A panel will explain the membership plans and your part in this progressive development.

PLACE Kimball - Legion Hall

DATE May 31st, 1955

TIME: 8 o'clock P. M.

MIDSTATE TELEPHONE COMPANY
Clifford Guernsey,

RURAL TELEPHONES

HEARINGS
BEFORE A
SUBCOMMITTEE OF THE
THE COMMITTEE ON AGRICULTURE
HOUSE OF REPRESENTATIVES
EIGHT

5 THE UNITED STATES OF
FIVE DOLLARS
A 53999401 A
UNITED STATES
ONE DOLLAR
WASHINGTON, D.C.
5888021B

ENMR
TELEPHONE
COOPERATIVE

TABLE OF CONTENTS

*A*fter just one evening of television commercials or a quick glance through the local paper, there is no denying that information and communications technologies permeate nearly every aspect of life. A multilane information superhighway runs across America, with even the most rural areas of the nation enjoying state-of-the-art communications services.

But that massive thoroughfare would not be possible today were it not for the merging of two small country lanes back in 1949. On October 28, President Harry Truman signed the telephone amendments to the Rural Electrification Act, joining the long, hard road of the independent telephone industry with the rugged trail the Rural Electrification Administration (REA) was blazing across rural America.

In the 50 years since that monumental day, both the independent industry and the REA (now the Rural Utilities Service (RUS)) have changed significantly, but without their work together, it is quite likely that no information superhighway—perhaps not even a strong communications infrastructure—would exist in rural America today.

In celebration of the accomplishments of this public-private partnership, the following pages document the past 50 years ... the struggles and the triumphs, the facts and the firsts, the innovators and the leaders. By bringing basic telephone service to rural Americans in the early years and assisting them in continuing to obtain state-of-the-art telecommunications today, the REA/RUS truly is ...

Builder of the Past, Architect of the Future

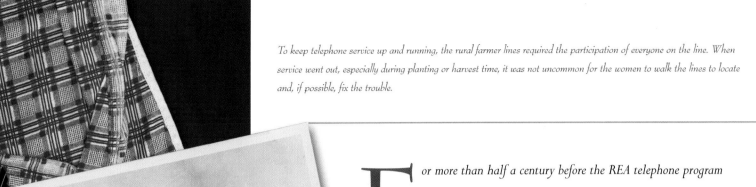

To keep telephone service up and running, the rural farmer lines required the participation of everyone on the line. When service went out, especially during planting or harvest time, it was not uncommon for the women to walk the lines to locate and, if possible, fix the trouble.

F or more than half a century before the REA telephone program came to be, telephone service existed in many rural areas, but for most, the service proved mediocre at best. From the invention of the instrument, to the days of the depression, early telephone service had some rough times. Yet the pioneering men and women who made such service possible formed the foundation upon which the REA telephone program later was built. And their entrepreneurial spirit, dedication to their communities, and pursuit of a better quality of life remained evident as the federal program developed and evolved. The story of the REA telephone program truly begins with the birth of …

The Independent Industry

Long before members of Congress and the industry even began talking about expanding the REA program to include telephone service, the U.S. Department of Agriculture (USDA) tried to lend farmers a hand in bringing telephone service to their farms. In 1922, the USDA published *Farmers' Bulletin No. 1245* titled Farmers Telephone Companies: Organization, Financing and Management. *The 30-page booklet provided statistics about rural telephone systems and offered detailed advice on forming a telephone organization, constructing the system, and managing the operations.*

In the Beginning

I t all started with Alexander Graham Bell, his assistant Thomas Watson, and those seven famous words: "Mr. Watson, come here, I want you." When Bell received his patent for the telephone— No. 174,465—in 1876, he launched what would become perhaps the most all-encompassing industry in the world.

While his invention met with some skepticism, many saw the potential of the instrument and began creating their own versions of the telephone. As Bell and his associates formed the Bell Telephone Company to provide service and distribute telephones, non-Bell—independent—telephone companies and manufacturers began operations as well.

The expiration of Bell's original patents in 1893 and 1894 cleared the way for the independent industry to truly move forward. In the next few years, thousands of telephone companies sprang up across the nation.

These non-Bell telephone systems took various forms. Some operated as small, family-owned businesses, often with the women running the switchboard and the men installing and repairing the lines and equipment. Others were commercial stock companies. But the majority were mutual organizations, the forerunners of cooperatives.

Stock mutuals charged member subscribers a regular fixed fee and usually operated their own switchboards. In contrast, members of "pure" mutuals, often called "club lines" or "farmer lines," each bore a share of the system expenses, did their own repairs, and paid no fixed service fees. Such mutuals, some of which served only a few stations, also did not have switchboards and often built a line into town to connect to the Bell company or another telephone system with a switchboard, bringing them the additional designation of "switcher lines." When one member of a farmer line called another member of the same line, the switchboard operator did not need to do anything.

Early rural telephone systems often had 20 or more subscribers connected to the same line. Obviously, service was far from private, but people generally did not mind because they viewed the telephone as a news medium as well as a communications tool. The operator, dubbed "central," provided a wealth of information, and with party lines, listening in on other people's conversations—called "rubbernecking"—was a common occurrence.

In the early days, subscribers purchased their own telephones—large, wall-mounted, magneto, crank instruments—from hardware or furniture stores or even mail order catalogs. Turning the crank on the telephone set sent out a current that would lower the associated drop on the switchboard, letting the operator know someone on that line wished to place a call. She would plug in the answering cord, ask what number the caller wanted, plug her ringing key into the drop of the person being called, and send out the appropriate rings. With multiple parties on the line, each had a coded ring consisting of a combination of short and long rings. If someone wished to call another person on the same party line, this did not involve the operator, but rather required the caller to use the crank to generate the other person's coded ring.

Early rural systems employed two types of construction—grounded and metallic. Grounded, one-wire systems had a single transmission wire and used the ground to complete the circuit. Sometimes farmers used existing barbed wire fencing as the conduit and strung wires from fence posts to dead trees. Service quality varied with the weather, as dry or frozen ground did not carry conversations well, and the presence of electrical circuits rendered grounded systems virtually useless. To connect to grounded circuits, customers needed magneto, or local-battery, telephones—the power to make the telephone work came from batteries within the instrument itself. Metallic, two-wire systems provided more reliable service, as they were not subject to the condition of the ground and could be constructed so as to reduce electrical circuit interference. Although the majority of rural subscribers had magneto telephones, metallic systems also could support common battery telephones, which were powered by large batteries at the telephone switching office.

From Chaos to the Commission

As the number of telephone systems grew, so did the complications. By the turn of the century, many communities had two or more telephone systems, and they were not interconnected. This required residents and businesses to subscribe to—and have a phone from—each company. Another problem was Bell-independent relations. The Bell telephone system, by then under the AT&T umbrella, refused to allow independents to interconnnect with its toll facilities and also began aggressively acquiring its independent competitors.

The duplication and other general concerns led to industry supervision by state regulatory commissions and, in 1910, the Interstate Commerce Commission. But disputes between AT&T and the independents continued. Federal officials began to threaten enforcement of antitrust laws or even legislation to nationalize telephone service.

Then, in 1912, Nathan C. Kingsbury, an AT&T vice president, wrote a letter to the U.S. Attorney General, which became known as the Kingsbury Commitment. He stated that AT&T would stop buying independents and allow them to obtain toll service for their customers via Bell system lines, as well as sell its stock in the Western Union Telegraph Corporation.

This agreement no longer governed, however, following the passage of the Communications Act of 1934, which established the Federal Communications Commission and gave it jurisdiction over interstate communications. The act also established the concept of universal service: all Americans, both urban and rural, are entitled to quality telephone service at reasonable rates.

A jumble of wires enters the building housing the old manual switchboard of East Ascension Telephone Company, Gonzales, Louisiana.

13

Depression, Disconnection, and Disrepair

By 1912, there were more than 32,000 rural telephone systems, and by 1917, the number had reached 53,000. Most were farmer lines with less than 100 telephones, and more than 51,000 had incomes of less than $5,000 per year. After World War I, the number of farmer lines continued to grow, with the overall number of rural systems reaching its peak of approximately 60,000 in 1927. But the systems were deteriorating—maintenance often was poor and subscribers frequently did not pay.

As the Great Depression fell upon the nation in the 1930s, the telephone situation progressed from bad to worse. Payment often came in the form of chickens, grain, or other farm products, if it came at all, and some people did manual labor for the telephone company to pay off their bills. Many subscribers discontinued telephone service, and revenues dropped significantly.

The majority of commercial companies and stock mutuals struggled along, but many farmer lines deteriorated right out of business. The advent of the Rural Electrification Administration electric loan program in 1935 accelerated the decline, as the rural electric lines often produced a humming noise on the grounded magneto telephone systems.

Fewer and fewer rural systems offered telephone service of any quality. Leaning poles, rusted wires, and broken equipment became the norm, and many systems' service was out as much as it was working. By the 1940s, it became clear that the rural magneto systems and one-wire grounded farmer line systems had to upgrade or go out of business. But this required capital—something the rural systems and their farmer subscribers did not have.

Broken-down equipment, such as that pictured here, was common throughout the nation. Testifying in favor of the REA telephone amendments in 1949, Billy Bryan, Cattle Electric Cooperative, Binger, Oklahoma, told the Senate committee of the dire situation. "Telephone lines are down in the grass. They are tied to fence posts. They are tied to black-jacks. They are stuck up on poles. They have no system. The phones are bad. … Did you ever see a fruit jar used as an insulator on a telephone line? We have lots of them …"

In the early 1930s, only about 10 percent of America's farms had electricity. This grim statistic served as the catalyst for the story of rural electrification, a story of rural people working with one another and the federal government to better their lives. Had this story not unfolded, the Rural Electrification Administration—and thus the REA telephone loan program—would not have come to be. The electric program set a strong example for the telephone program, as the agency's early engineers and the nation's rural electric cooperatives undertook the task of ...

Lighting the Land

As the REA began to make loans and electric cooperatives began their operations, farmers throughout the nation celebrated as the lights came on.

Then-New York Governor Franklin Roosevelt works with his personal assistant, V. Warwick Halsey, at his cottage in Warm Springs, Georgia, the location he later called the "birthplace of REA." Speaking at the dedication of Lamar Electric Membership Cooperative, Barnesville, Georgia, in 1938, Roosevelt told how 14 years earlier he went to Warm Springs to recover from a polio attack, and when he received his first month's electric bill, found it to be four times what he paid for electric in Hyde Park, New York. "That started my long study of public utility charges for electric current and the whole subject of getting electricity into farm homes."

The REA is Born

Senator George Norris (R-NE) and Representative Sam Rayburn (D-TX) introduced the measures that became the Rural Electrification Act of 1936.

When President Franklin Roosevelt took office on March 4, 1933, the country was still reeling from the Great Depression. Looking to get the nation back on its feet, Roosevelt launched his New Deal initiatives and during the now famous First 100 Days of his administration, took several monumental actions, including the creation of the Tennessee Valley Authority.

Later that year, Morris Cooke, a Philadelphia engineer working on rural power issues in Pennsylvania and then New York, came to Washington, D.C., to join the Roosevelt administration. He urged the President to launch a rural electrification program, but while Roosevelt supported the idea, the many New Deal projects took precedence.

In 1934, Cooke wrote a detailed memo to Interior Secretary Harold Ickes outlining the feasibility of a national rural electrification program. This document, along with pressure from various farm groups and the need to address continued unemployment, brought the matter to the forefront. On May 11, 1935, Roosevelt signed Executive Order 7037, which created the Rural Electrification Administration (REA) under the authority of the Emergency Relief Appropriation Act of 1935. Cooke became the first REA administrator and served until early 1937.

The REA received $100,000 to carry out its rural electrification efforts, but as an unemployment relief agency, it had to spend at least 25 percent of the funds for labor and hire 90 percent of that labor from relief rolls. It didn't take long for Cooke and the others to realize that the REA did not fit the relief program mold because installing electric facilities required skilled labor. Roosevelt agreed to eliminate the relief role requirement for the REA and established it as a lending agency.

Because REA funding remained tied to an emergency appropriation act, many of its supporters feared its mission would not be accomplished. Thus, in early 1936, two strong REA advocates on Capitol Hill—Senator George Norris (R-NE) and Representative Sam Rayburn (D-TX)—introduced measures to establish the REA as an independent agency receiving regular appropriations. Congress passed the legislation, and Roosevelt signed the Rural Electrification Act on May 20, 1936.

The REA now stood as a full-fledged lending agency making interest-bearing, self-liquidating loans to organizations that would bring electric service to America's farmers. REA loans carried an interest rate that fluctuated with the government's cost of money and a repayment period of 25 years. The 1936 act authorized the agency for 10 years.

Realizing an understanding of the true power of electricity on the farm was key to the successful expansion of the REA electric program—and to the financial stability of its borrowers—the REA decided it must spread the word. In late 1938, the REA took its show on the road, launching a traveling farm equipment show, often referred to as the REA Circus. With a tent that seated 1,000 and trucks and trailers filled with electric appliances and equipment, REA staff members set out across the country. The crew included technical and information specialists, as well as home economists who demonstrated household appliances. The REA's Dan Teare, pictured here giving a demonstration, was known as the circus' "ringmaster." As the show's popularity grew, appliance manufacturers joined in as well, setting up their displays near the REA big top. The tour continued until late 1941 when it folded due to the war. During its run, the circus reached nearly one million farmers in more than 20 states.

Cooperatives Make the Connections

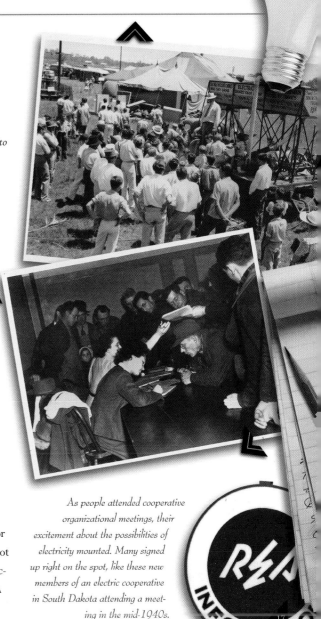

As people attended cooperative organizational meetings, their excitement about the possibilities of electricity mounted. Many signed up right on the spot, like these new members of an electric cooperative in South Dakota attending a meeting in the mid-1940s.

At first the REA intended to make loans to large commercial electric companies, giving them the financing to extend service to rural areas. But these companies showed little interest, and it soon became clear that it would be cooperatives—groups of local people working together on their own behalf—that took the lead in electrifying rural America.

As they learned about the REA program and its mission, rural citizens began to form cooperatives. Community leaders held informational meetings, began planning the cooperative and electric system, and undertook the task of getting people to join the cooperative. Participation required a $5 payment, plus a realization that using electricity would mandate wiring the home and purchasing the necessary fixtures, appliances, and equipment. Cooperative leaders canvassed residents, explaining the organization's goals and the countless benefits of electricity.

Unfortunately, many newly formed or soon-to-be-formed cooperatives faced "spite lines" or "cream skimming" by the local power company. The commercial operation would build a line into the area and hook up the businesses and easy-to-reach farms—the very subscribers the cooperative needed to make its system feasible. This weakened and, in many cases, destroyed the fledgling cooperatives. But despite the obstacles, rural electric cooperatives quickly sprang up throughout the nation. By 1937, the REA had 266 electric borrowers, nearly all of which were local cooperatives.

While rural Americans were busy establishing their cooperatives, the REA was busy with its own task of figuring out the best way to bring electricity to rural customers at reasonable rates. Because of the depression, the REA was able to hire top engineers, and these men became rural service innovators—a designation that would be a hallmark of the REA for years to come. They developed standards and new construction methods specific to rural electrification.

By now, the REA pattern had been established. The five-part program approach—filling a real need for those it intended to serve; promoting the general welfare of the nation; providing self-liquidating loans, not grants; providing service to all persons within a given area; and operating within the framework of democracy and free enterprise—became a model for future endeavors both here and abroad, including the REA telephone program which soon would follow.

17

Service for Everyone— No Matter How Far

Two more legislative actions affecting the REA's standing as a government agency occurred during its initial decade of operation. First, on July 7, 1939, under the Government Reorganization Act of 1939, the REA became part of the U.S. Department of Agriculture and was no longer an independent agency. When the agency lost its independent status, REA Administrator John Carmody stepped down, and Harry Slattery took his place.

Then, as the REA's 10-year loan authority approached its expiration, Representative Stephen Pace (D-GA) introduced a measure that not only extended the agency's loan authority indefinitely, but also made several significant changes to the REA program. The Department of Agriculture Organic Act (the Pace Act), signed into law in September 1944, established a fixed interest rate of 2 percent for REA loans and extended the repayment period to 35 years.

But most importantly, the measure established the concept of area coverage—the practice of extending service to everyone in a given area who wants it regardless of how far they live from the main facilities; of considering the feasibility of a project as a whole, not based on the cost of individual extensions—and made area coverage a condition for receiving an REA loan. The area coverage principle proved key to bringing both electric and telephone service to rural America.

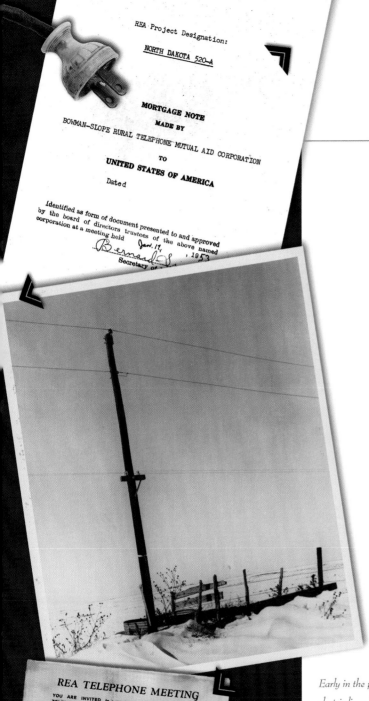

REA Project Designation:

NORTH DAKOTA 520-A

MORTGAGE NOTE
MADE BY
BOWMAN-SLOPE RURAL TELEPHONE MUTUAL AID CORPORATION
TO
UNITED STATES OF AMERICA

Dated

Identified as form of document presented to and approved
by the board of directors trustees of the above named
corporation at a meeting held Jan. 17, 1953

Bernard S.
Secretary of

REA TELEPHONE MEETING

YOU ARE INVITED to a meeting sponsored by the MIDSTATE
TELEPHONE COMPANY for the purpose of explaining plans for
providing your area with a modern dial telephone system. A panel
will explain the membership plans and your part in this progressive
development.

PLACE *Kimball-Legion Hall*
DATE *May 31st*, 1955
TIME: *8* o'clock *P.* M.

MIDSTATE TELEPHONE COMPANY
Clifford Guernsey,
Manager

By the late 1940s, the war was over and Americans were anxious to get on with their lives. But the war-time economy and the growth of rural electric systems had taken their toll on rural telephone systems—in 1949, only about one in three farms had telephone service, and a significant portion of that service was 20-party. That year, Congress passed amendments to the Rural Electrification Act, creating the telephone loan program and setting in motion the public-private partnership that would change the face of communications in rural America. The program's pioneers worked through the many details and decisions of initiating loans and soon the REA was …

Funding Phone Service

Early in the program, REA engineers considered the possibility of transmitting telephone signals via existing rural electric lines. The necessary technology existed, but the power line carrier equipment proved too expensive and also could not provide selective ringing for party lines. The REA did, however, encourage the joint use of electric poles for telephone wiring, as pictured here.

President Harry Truman (seated) signs the legislation adding the telephone loan program provisions to the Rural Electrification Act. Attending the signing ceremony are (from left) Kit Haynes, National Council of Farm Cooperatives; Benton Stong, National Farmers Union; Secretary of Agriculture Charles Brannan; Tom Duncan, American Federation of Labor; Clyde Ellis and Clark McWhorter, National Rural Electric Cooperative Association; REA Administrator Claude Wickard; Roger Fleming, American Farm Bureau; and Dr. J.T. Sanders, National Grange.

Telephone Loans Become Law

E*ven in the late 1930s, there was talk of the government's potential role in helping* to bring telephone service to rural areas. Writing to an editor in North Carolina in 1939, REA Administrator John Carmody stated: "Personally, I have long felt that there was a real opportunity for constructive assistance to rural people in the idea of Federal financing of farm telephone lines. It seems to me that rural people have just as much right to up-to-date communication as they have to modern power. There's no question in my mind but that Government assistance will be required if the job is ever to be completed." At that time, fewer farms had telephone service than did in 1920, and the situation showed little hope for improvement.

A few years later, those on Capitol Hill began to look at the matter. In late 1944, Senator Lister Hill (D-AL) introduced a bill calling for the formation of the Rural Telephone Administration, a new agency modeled after the REA. He credited Gordon Persons, president of the Alabama Public Service Commission, with proposing the idea of a federal rural telephone program.

The following year, both Hill and Representative W.R. "Bob" Poage (D-TX) introduced measures to establish a rural telephone loan program, with Poage's proposal calling for the REA to make the loans. But Congress did not take up either bill. Hill and Poage re-introduced legislation each successive Congress until action finally came in 1949, when a subcommittee of the House Agriculture Committee held hearings in February, and the Senate Committee on Agriculture and Forestry held hearings in June and August.

Members of the Truman administration, including REA Administrator Claude Wickard, strongly favored passage of the legislation. They and other proponents testified to the dire need for a telephone loan program. The REA cited letters and telegrams from rural people complaining about many existing telephone companies—how they had been promised service for years, refused service outright, or given terribly inferior service. Individual farmers shared their stories, and several persons representing existing mutual or cooperative telephone systems stressed their desire to have government loans available.

The National Rural Electric Cooperative Association (NRECA), through the testimony of Executive Manager Clyde Ellis, stood as one of the bills' strongest supporters. The association's Communications Committee focused on the expansion of rural telephone service, and at several of its annual meetings during the 1940s, the NRECA passed resolutions addressing the matter. Also supporting the measures were the National Grange and the National Farmers Union.

Opposition to the bills came from the Bell companies and many commercial independent telephone companies who claimed progress in rural telephony to be sufficient without the government's involvement. The United States Independent Telephone Association (USITA) called for using the Reconstruction Finance Corporation, instead of the REA, for telephone financing. Many opponents also expressed concerns about duplication of facilities, unfair competition, and government operation of telephone systems.

The debate brought some modifications to the bills, specifically the elimination of public bodies from the list of eligible entities and a preference for existing telephone organizations to receive loans. Revised measures passed the House in July and the Senate in September. Then, on October 28, 1949, President Harry Truman signed the telephone amendments to the Rural Electrification Act into law. The REA now could make self-liquidating loans at 2 percent for up to 35 years for the extension and improvement of rural telephone service.

Rural telephone amendment authors Representative W.R. "Bob" Poage (D-TX) (left) and Senator Lister Hill (D-AL) (right) join REA Administrator Claude Wickard (center) as he approves the first REA telephone loan.

RURAL TELEPHONES

HEARING
BEFORE A
SUBCOMMITTEE OF THE
COMMITTEE ON AGRICULTURE
HOUSE OF REPRESENTATIVES
EIGHTY-FIRST CONGRESS
FIRST SESSION

FEBRUARY 14, 15, AND 16, 1949

Serial B

The Core Components

Mrs. Cricket Taylor (right) tells Mrs. Fred Baldridge about the Eastern New Mexico Rural Telephone Cooperative, Clovis, New Mexico, and inquires as to her interest in obtaining membership and service. Subscriber surveys assisted potential REA borrowers in more accurately planning their financing needs.

T he new law set forth the basic parameters for the REA to carry out the congressional mandate of assuring "the availability of adequate telephone service to the widest practicable number of rural users of such service."

In terms of eligibility, the law required that during the first year, preference be given to applications from telephone systems already in operation. The government believed the existing companies were key to the program's success, because most rural areas already were assigned to them by state regulatory bodies, and thus encouraged them to extend their service. But in areas where such companies did not exist or were unable or unwilling to extend their service, the REA supported the formation of new telephone systems.

While the main purpose of the telephone loans was the extension and improvement of service in rural areas—areas with a population of fewer than 1,500—borrowers could use a portion of the funds to purchase, construct, or improve facilities in non-rural areas if doing so proved necessary for providing adequate service in the rural areas. Borrowers also could use up to 40 percent of a loan to refinance outstanding debt if, again, it helped them better serve rural areas. Additionally, the law allowed borrowers to use loan funds for the acquisition of existing lines or systems, with some limitations.

The law charged the REA administrator with determining that each loan was indeed feasible and, when states did not provide certification, with determining that the proposed system did not duplicate the facilities of another system providing reasonably adequate service. The REA made loans for the construction or improvement of an entire system or just a portion of a system, and the agency set no limit on the loan amount, so long as it appeared economically feasible.

The REA secured its loans by acquiring the first lien on a borrower's total system—including everything the borrower owned prior to the loan and all assets it acquired thereafter. The terms of the REA mortgage additionally gave the agency control over matters such as the payment of dividends and salary increases. Borrowers paid interest quarterly, and principal payments generally did not begin until three years after the loan date so as to give borrowers the opportunity to complete construction and begin taking in revenues.

Decisions, Decisions, Decisions

Tell Your Friends

Plan Now to Attend One of these
MIDSTATE TELEPHONE COMPANY
EQUITY MEETINGS

Tuesday-May 31
Legion Hall-Kimball 8:00 p.m.
FRED HOUDA WILL ACT AS MODERATOR — FREE COFFE & DOUGHNUTS SERVED

Friday-June 3
City Hall-Pukwana 8:00 p.m.

HERE IS YOUR OPPORTUNITY TO FIND THE CORRECT ANSWER TO ANY QUESTIONS
MAY HAVE OF THE FUTURE PLANS AND PURPOSES OF THE MIDSTATE ORGANIZA
A PANEL WILL ANSWER YOUR QUESTIONS FROM THE FLOOR.

Applications For Membership Will Be Accepte

E ven with the language of the law, the staff of the new REA telephone loan program faced many unanswered questions as they began establishing policies and loan procedures to get the program under way. They found the telephone problems of most rural areas to be quite complex and also found telephone service to be more complicated than electric service.

One of the REA's most significant decisions was that of not simply patching up existing magneto systems, but rather financing only dial service and requiring that it be eight-party service or better so it could provide full selective ringing. Instead of the earlier system of coded, multiparty ringing, selective ringing meant only the person being called heard his or her telephone ring. Borrowers could, however, retain existing common battery facilities within their systems in areas where that proved most economical.

And, as in the electric program, borrowers had to adhere to the REA's area coverage principle. This meant planning and constructing the system as a whole to make service available to all people who wanted it, regardless of whether the installation and operation of a particular line proved profitable. As part of the area coverage requirement, borrowers had to offer uniform rates for each grade of service, no matter where an individual lived, with no zone or mileage charges.

One of the agency's early decisions that brought much contention in the industry was the requirement that borrowers provide equity funding to supplement their REA loans. The REA deemed such a contribution necessary for loan security, citing the high probability of customers' discontinuing service in times of financial difficulty. Thus, the REA telephone program did not make 100 percent loans, but rather loans for approximately 50 to 90 percent of the project costs.

The exact equity amount depended upon the individual system and the value of any existing telephone facilities, but ranged from 20 to 30 percent equity for commercial companies and $25 to $50 per subscriber for cooperatives (the equivalent of nearly $200 to $400 today). This was quite a bit more than the $5 per subscriber the REA required electric cooperatives to collect. The controversy centered on the equity level for telephone cooperatives, as well as what the industry perceived as the REA's stricter application of the equity requirements to cooperatives than to commercial systems.

Other matters the REA had to address included finding ways to estimate the market for service in each area, developing loan feasibility guidelines, and establishing construction and equipment standards. After about two months of wrestling with program details, the REA began distributing telephone loan applications on December 22, 1949.

Retired Bell and AT&T employees, as well as individuals from the independent industry, came to work in the REA telephone program and share their expertise. Even in 1950, many of them had 25 or more years of experience. Among the talented independent industry experts coming to the REA and making an enormous contribution to the telephone program was Daniel Corman, manager of South Central Rural Telephone Cooperative, Glasgow, Kentucky, who acted as consultant to the REA administrator from 1949 to 1952. At the 1975 National Telephone Cooperative Association (NTCA) Annual Meeting, Corman (center) is recognized for his 72 years of service to the industry. Presenting the award on behalf of NTCA is REA Administrator David Hamil (left). Corman is joined by his nephew, William Corman of Southland Telephone Company, Atmore, Alabama.

Ready to Roll

T he telephone program started with three staff people—an engineer, an information specialist, and a secretary—and built from there. Assembling the technical staff proved a difficult job in the post-war period, because telephone companies and equipment manufacturers also were hiring. The REA transferred people from its electric program staff and hired from within the telephone industry. The agency also secured several retired Bell and AT&T employees as consultants. By June 30, 1950, the telephone program had 142 full-time employees.

In the six weeks after the telephone amendment became law, REA received more than 700 requests for information. About 600 of these came from existing systems wanting to modernize or extend their service, while the remainder came from newly forming mutuals or cooperatives. Within three months of the law's enactment, REA had received 1,117 loan inquiries from potential applicants in 44 states. The REA established a special Telephone Information Unit to handle requests, and by the end of fiscal year 1950, had received 486 loan applications from 41 states and Alaska.

The REA's electric cooperatives made an all-out effort to help get telephone service to rural areas, as they saw many advantages to their customers' having a telephone. They publicized the telephone cooperatives' organizational meetings in their newsletters, helped new telephone organizations prepare their REA loan applications, and provided membership lists for recruiting purposes. In late 1949, REA Administrator Claude Wickard wrote a letter to electric borrowers encouraging such efforts.

Within less than a year, the REA telephone loan program was well on its way to offering much-needed financing, as well as technical and management assistance, to rural telephone systems.

Throughout the nation, REA electric cooperatives spearheaded the efforts to get telephone service in their areas. South Plains Electric Cooperative played a major role in the development of South Plains Rural Telephone Cooperative, Lubbock, Texas, assisting with everything from legal advice to subscriber sign-up efforts. Here Mr. J.G. Walker of South Plains Rural Telephone Cooperative talks to Mrs. W.E. Burnett about joining.

Creating Cooperatives

A s rural telephone systems began looking at REA financing—and the REA began looking at potential rural borrowers—it became obvious that many of the mutuals and farmer lines were just too small to handle the upgrade to dial service. They basically had no choice but to sell or consolidate with one another. Numerous cooperatives, and even some commercial operations, formed as a result of such mergers.

A prime example of this approach took place in Thompson, Iowa, where in late November 1949, the Winnebago Rural Electric Cooperative Association invited 57 local mutual telephone companies to attend a discussion meeting. The 107 representatives from more than 20 mutual companies who attended all agreed to poll their members concerning a possible consolidation.

When nearly all reported their members' desire to merge, the electric cooperative board agreed to organize and operate the Winnebago Cooperative Telephone Association. Electric cooperative manager Glenn Bergland took on the management of the telephone cooperative as well. He and other representatives attended several meetings per day during the next few months as they met with each mutual about becoming part of the cooperative.

More and more systems joined the group until in June 1950, the cooperative—then consisting of the combined members of 43 mutuals—successfully applied for an REA loan. Winnebago received a $552,000 loan allocation in October 1950, and it became a national model for other newly organized cooperative groups, as well as for electric cooperatives expanding into the telephone business. The Winnebago cooperative continued to add additional telephone systems to its operations in the years that followed.

But whether forming a cooperative from existing mutual systems or launching the operation from scratch, rural citizens often found the start-up process to be a challenging task. Organizers worked tirelessly holding informational meetings, promoting the forming cooperative in the local media, and going door to door signing up member subscribers.

Because the subscribers owned a portion of the company, no REA loan funds could be used to pay for rights of way on members' land, and getting the necessary right-of-way certificates often proved a difficult process. One cooperative employee recalls being chased by a woman with a shovel when he suggested placing telephone facilities on her land.

In December 1950, Manager Jim Wright (right) signs up Mr. and Mrs. E.J. Lassetter for membership in Cap Rock Telephone Cooperative, Spur, Texas. Not only did the Lassetters join, but they also leased the cooperative the land for its headquarters building—for 99 years for $1.

And like their fellow rural electric cooperatives, these new telephone systems were not immune to competitive pressures. Often a Bell company or larger independent would extend service into the cooperative's territory, connecting only the most profitable customers and rendering the cooperative's operations unfeasible. The high per-subscriber equity requirement presented a challenge as well.

The REA field staff offered a great deal of assistance, attending the organizational meetings to answer questions and helping the rural systems complete their REA loan applications. Once the REA received an application, it assigned a team of REA operations and engineering people to further assist the system.

Today, approximately 260 telephone cooperatives operate in the United States, and they do so under the same principles as those early cooperatives—open membership; one member, one vote; non-profit operations; and returning a percentage of margins earned to members in the form of capital credits. The cooperative determines each member's capital credits based on the total dollar volume of business that member has done with the cooperative and then pays or "retires" the capital credits periodically based on a formula determined by the cooperative board. This process is an essential part of member ownership in the cooperative.

Before obtaining an REA loan and beginning operations, rural telephone systems had to organize and determine their specific communications needs at community meetings such as this gathering of the fledgling Citizens Telephone Cooperative, Floyd, Virginia. Often REA representatives assisted in these efforts.

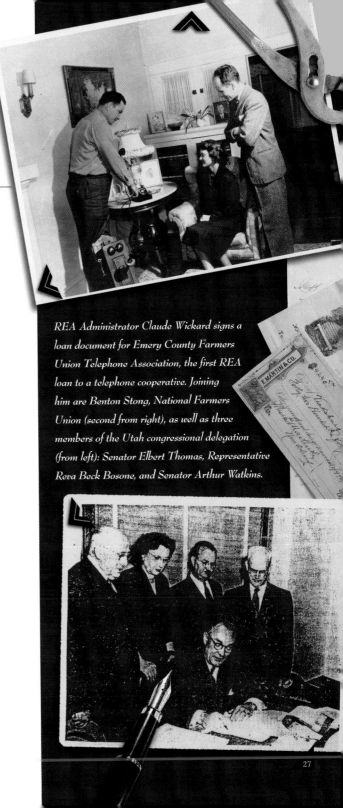

Florala Telephone Company Manager Lloyd G. Vaughan (left) shows Elaine and Ewell Clark how to use their newly installed dial telephone. Florala received the first REA telephone loan.

Funds Begin to Flow

On February 24, 1950, Florala Telephone Company, Florala, Alabama, became the first telephone system to receive an REA loan. The telephone company used the $243,000 loan to rebuild its existing system and to add 130 miles of new telephone line. Of that, 80 miles were added by sharing pole space with three REA electric cooperatives.

The construction allowed Florala to better serve its 447 current subscribers and extend service to 1,053 more. Plus, the network backbone positioned the company to easily add the remaining 960 subscribers in its next expansion and thus achieve area coverage.

On approving the first REA telephone loan, Administrator Claude Wickard stated, "This marks the first application of the REA pattern outside the field of rural electric service."

Less than two months later, on April 19, 1950, Emery County Farmers Union Telephone Association, Orangeville, Utah, became the first cooperative to receive REA financing. It was the fifth loan allocation announced by the agency. The cooperative used the $222,000 loan to provide an area coverage system in a valley area virtually without service. It installed a complete dial system including four central office units, 177 miles of pole line, and a headquarters building. When the cooperative initiated service in 1953, it was serving 584 members.

Emery also claims the distinction of submitting the first REA telephone loan application, under its original name, the Castle Valley Telephone Association. It had submitted the application in early 1950 on the advice of the National Farmers Union, which not only was working to organize farmers in Utah, but also had actively lobbied for the passage of the REA telephone amendments in Washington, D.C. Thus, when incorporating as a non-profit corporation in late 1950, the cooperative members included "Farmers Union" in the name in recognition of the work that group had done on behalf of rural telephony.

By the end of fiscal year 1950, the REA had made 17 loan allocations in 15 states. The partnership of rural people, industry suppliers, and the government had successfully begun its mission of bringing telephone service to rural America.

REA Administrator Claude Wickard signs a loan document for Emery County Farmers Union Telephone Association, the first REA loan to a telephone cooperative. Joining him are Benton Stong, National Farmers Union (second from right), as well as three members of the Utah congressional delegation (from left): Senator Elbert Thomas, Representative Reva Beck Bosone, and Senator Arthur Watkins.

UNITED STATES DEPARTMENT OF AGRICULTURE
Rural Electrification Administration

Virginia Farm Couple to Get Call from President on First REA Phone

The city girl who married her country boy sweetheart and her husband, an Air Corps veteran, today are in a fair way to find themselves one of the most talked about rural couples in the United States.

They are Mr. and Mrs. Eugene Dickinson, RR1, Fredericksburg, Virginia, the first rural family in the country to get telephone service as a direct result of the new REA telephone loan program.

As a result, they will receive a call from President Truman in the ceremonies to be held on "REA Telephone Day" at the Fredericksburg Agricultural Fair on Wednesday, September 20.

Since the Fredericksburg and Wilderness Telephone Company installed the line and a dial phone to give the couple their first telephone during their five and a half years of married life, the Dickinsons ...

that it's just one thing after another, most ...

The modern luxury trail ...
by reporters ...

The First of Many

Wednesday, September 20, 1950. It was a celebration day that touched many more Americans than those living in the small Virginia community in which it was held. On that day, Fredericksburg and Wilderness Telephone Company, Chancellor, Virginia, marked the cutover of the first REA-financed facilities into service, celebrating REA Telephone Day at the Fredericksburg Agricultural Fair.

As part of the day's events, President Harry Truman placed a telephone call to Mr. and Mrs. Eugene Dickinson, the first farm family in the nation to receive service as a result of the REA telephone loan program. Many REA representatives, members of Congress, and other dignitaries, joined in the festivities.

Within 30 days of the new dial phones replacing the hand-crank models, the 150-line telephone company reported its average daily call volume had increased from 250 to 1,500. Subscribers had gone from making less than two calls per day to making up to 10. Not only was the service quality better, but they no longer had to pay an extra toll each time they called into Fredericksburg.

Customer accounts recorded in the December 1950/January 1951 issue of the REA's *Rural Electrification News* magazine clearly showed that the improved service was not a luxury for the citizens of Chancellor, but rather a necessity. Farmer Andrew Seay, Acting Post Mistress Lucille McGhee, Chancellor Elementary School Principal Alfred Moyse, and community physician Dr. Henry Bernstein all recounted the ways in which their new telephones already had made their work lives more productive and increased the overall safety and well being of those in the community.

The new phones did have their social appeal as well. As Mrs. Dickinson stated, "I didn't know I had so many friends. I just love the phone."

A large crowd gathers in Chancellor, Virginia, for the ceremony marking the cutover of the first REA-financed telephone facilities by Fredericksburg and Wilderness Telephone Company.

Facts and Figures: *REA/RUS Firsts*

Through the years, the REA/RUS telephone loan program has recorded many milestones. The following are among the agency's most significant "firsts."

2/24/50 ...... **First** telephone loan—Florala Telephone Company, Florala, Alabama

4/19/50 ...... **First** telephone loan to a cooperative—Emery County Farmers Union Telephone Association, Orangeville, Utah

9/20/50 ...... **First** cutover of REA-financed facilities—Fredericksburg and Wilderness Telephone Company, Chancellor, Virginia

8/23/52 ...... **First** cutover of REA-financed facilities by a cooperative—Farmers Mutual Telephone Company, Shellsburg, Iowa

9/52 .... **First** loan repayment made—Fredericksburg and Wilderness Telephone Company, Chancellor, Virginia

11/8/52 ...... **First** entirely new exchange built by a new telephone enterprise under REA went into service—Eastern New Mexico Rural Telephone Cooperative, Clovis, New Mexico

10/31/63 .... **First** loan to an REA borrower to provide all one-party service—Dunnell Telephone Company, Dunnell, Minnesota

11/4/63 ...... Billionth dollar in loan funds approved—Norway Rural Telephone Company, Kanawha, Iowa

12/71 .... **First** Rural Telephone Bank Class C stock certificate issued—Tuolumne Telephone Company, Tuolumne, California

1/21/72 ...... **First** Rural Telephone Bank loans—Plant Telephone and Power Company, Tifton, Georgia; Mutual Telephone Company, Sioux Center, Iowa; Pioneer Telephone Association, Ulysses, Kansas; Randolph Telephone Company, Liberty, North Carolina; Eastex Telephone Cooperative, Henderson, Texas

12/11/74 .... **First** guaranteed loan commitments—Doniphan Telephone Company, Doniphan, Missouri; St. Joseph Telephone and Telegraph Company, Port St. Joe, Florida; West Virginia Telephone Company, St. Marys, West Virginia

4/10/75 ...... **First** guaranteed loan advance of funds—Doniphan Telephone Company, Doniphan, Missouri

3/19/76 ...... **First** loan to an Indian tribe and first loan to a public body—Cheyenne River Sioux Tribe Telephone Authority, Eagle Butte, South Dakota

5/89 .... **First** Rural Economic Development Loan and Grant recipients selected, including two telephone borrowers: The Pine Telephone Company, Wright City, Oklahoma, and Cheyenne River Sioux Tribe Telephone Authority, Eagle Butte, South Dakota

9/93 .... **First** Distance Learning and Telemedicine Grant recipients selected

10/97 .... **First** Distance Learning and Telemedicine Loan recipients selected

Despite the abundance of information inquiries and the excitement of the "firsts," the REA telephone program encountered some difficulties during the early years. The Korean War brought a materials shortage, and borrowers faced delays in central office equipment deliveries, as well as slow subscriber sign-up rates. Yet the REA staff forged ahead with their mission and undertook the initial engineering and standards-setting tasks that established the base not only of the REA telephone program, but of the entire independent telephone industry. From manufacturer research contracts and prototype equipment to engineering innovations and new construction techniques, the 1950s truly proved to be the …

Decade of Development

Barry Lavett of Eastern New Mexico Rural Telephone Cooperative, Clovis, New Mexico, stops to do a routine check at an unattended dial office. Throughout the nation, rural telephone systems constructed such buildings to house the switching equipment needed to serve their rural exchanges.

Charting the Course

As the REA staff set about establishing the framework for the new telephone program, they first looked at existing telephone technology and system design to determine what would work for rural telephone systems. The answer, unfortunately, proved to be "very little." In most instances, conventional equipment and construction methods were not feasible solutions for initiating or extending service in sparsely populated rural areas. Thus, the REA engineers began the daunting task of developing entirely new concepts of telephone engineering, manufacturing, and construction specifically for rural telephony.

To ensure the most reliable and cost-effective systems, the REA wanted its borrowers to use only equipment and supplies meeting acceptable standards. But the REA found that for the most part, no industry standards existed. To allow borrowers to get started on their systems, REA engineers quickly assembled interim standards and equipment specifications, as well as a tentative list of acceptable equipment and materials. They then went to work developing standards and specifications specifically designed to address the unique characteristics and problems of rural areas. Incidentally, many of these REA-developed standards became the industry standards.

But the process was not one of simply writing and issuing standards and specifications—it required a great deal of consultation work, as well as some sales and marketing effort. Manufacturers did not make equipment for the rural telephony market and had to be convinced sufficient demand existed before they would commit the time and resources. The REA staff visited manufacturers and attended industry conferences, explaining the REA program and the many benefits of rural telephony.

To encourage manufacturers to produce new equipment geared to the needs of rural telephone systems, the REA decided to initiate contractual research projects through which it would partially fund manufacturers' development of prototype equipment. Then, for testing the new equipment, the REA selected one or more of its borrowers to conduct a field trial. This practice of field testing equipment in an actual borrower system prior to adding it to the REA list of approved materials remains in effect today. The REA also worked closely with suppliers, as well as the National Bureau of Standards (now the National Institute of Standards and Technology), the U.S. Army, and other government entities to develop and implement testing procedures.

In early 1952, the REA signed its first two research contracts with telephone equipment manufacturers—one with Budelman Radio Corporation for the development of multiparty subscriber carrier equipment and the other with Raytheon Manufacturing Company for the development of low-cost point-to-point microwave radio. Many other such contracts would follow in the years ahead.

Mr. and Mrs. J.O. Jackson of Fort Morgan, Alabama, use their new radio telephone service provided by Gulf Telephone Company, Foley, Alabama. In 1052, Gulf was one of the first telephone systems to use radio telephone equipment to bring service to isolated areas for which pole line service proved impractical. The subscriber's line terminated in a radio station that was both a transmitter and receiver, which communicated with a similar station at the central office in Foley.

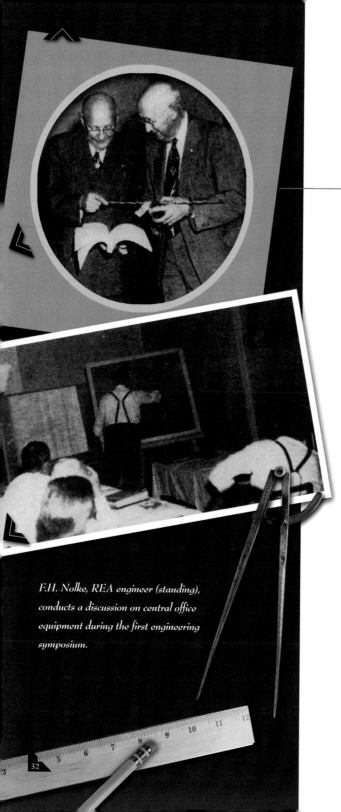

Consulting engineers C.E. Zahm and L.P. Allen talk about line wire ties during the first REA engineering symposium held in Cheyenne, Wyoming.

F.H. Nolke, REA engineer (standing), conducts a discussion on central office equipment during the first engineering symposium.

Training Others to Help

From crossarm construction to cable cladding, every engineering and construction idea fell under the scrutiny of whether it would be cost effective, not only in terms of low installation costs, but also reduced maintenance. And as the REA staff continually searched for better methods and materials and worked tirelessly to lower the costs of rural construction, they pioneered many innovations in equipment design and construction techniques that benefited the entire telephone industry.

The REA's work resulted in a standard of quality that surpassed that of non-borrower systems and led to borrowers' providing more sophisticated services than those offered in the rural areas served by large telephone companies. The agency's standards, combined with its list of approved materials, produced uniform system designs and thus gave borrowers increased purchasing power.

The agency's extensive work on rural telephony concepts also led to its publication of the *Telephone Engineering and Construction Manual* in 1951. This loose-leaf binder provided borrowers and their engineers a wealth of technical information on every phase of rural system development and became an indispensable reference tool. The REA continued issuing updates to the manual until the mid-1990s. The agency still maintains and regularly updates its "List of Materials Acceptable for Use on Telecommunications Systems of RUS Borrowers," first published in fiscal year 1951, with the primary distribution now being via the RUS web site.

Not only were appropriate standards, practices, and equipment lacking in the early years, but so were engineering firms qualified to design the rural systems the REA financed. To combat this problem, the REA developed a series of symposia designed to generate interest in the growing rural market and train the engineering firms so that as the number of borrowers increased, there would be qualified consulting engineering firms to assist them—not just the REA engineers.

The REA held its first two symposia in November 1952—one in Washington, D.C., and the other in Cheyenne, Wyoming. More than 80 engineers from 48 firms attended the events and became the basis of a strong field of engineers specializing in rural telephony. These engineering training sessions have continued throughout the program's history, and today the Engineering and Management Symposia is held every two years. Attendees include telephone company personnel in addition to engineering firm representatives.

Day-to-Day Details

The REA's work with its rural telephone borrowers encompassed not only the engineering and construction aspects of their systems, but the operations and management portions as well. In 1953, the REA published its *Telephone Operations Manual*, a loose-leaf binder to assist borrowers in establishing effective processes and procedures in five operational areas—general, plant, traffic, commercial, and accounting. The manual provided much-needed assistance to the small systems, where a few people often had responsibility for everything. Nearly 50 years later, the REA continues to issue periodic updates to the manual.

In addition to its flagship manuals, the REA produced a wealth of publications covering all aspects of rural telephone operations, each designed to guide borrowers toward becoming more self-sufficient. The agency offered a variety of instructional courses and a safety program, as well as aided borrowers in their service and equipment marketing efforts.

Many borrowers required assistance in negotiating appropriate connecting agreements with neighboring telephone companies for toll, operator, or extended area service. Also, when needed, the REA helped borrowers prepare comprehensive management studies and find solutions to management problems.

The REA acted as the claimant agency, on behalf of borrowers, for the materials control programs during the Korean War and provided technical assistance to those borrowers providing specialized communications facilities for Department of Defense installations and projects.

Looking to make its work with borrowers even more effective, the REA organized a Telephone Advisory Committee in 1954. The committee included representatives from REA borrower systems, as well as other segments of the rural telephone industry. The group met periodically in Washington, D.C., and after each gathering, made recommendations to the REA administrator about the program.

Another focus of the REA's marketing assistance came in promoting the sale of extension phones and other telephone accessories. Merchants & Farmers Telephone Company, Montpelier, Virginia, put the REA sales suggestions into action—Combination Man Robert Moody installs a barnyard gong for a dairy farmer so he can hear his telephone above the noise in the barn.

In the general store scene, citizens discuss the need for better telephone service.

Phyllis Sprecker, the Scarville, Iowa, housewife who starred in The Telephone and the Farmer, *joins a fellow actor in shooting an outdoor scene for the REA promotional film.*

The Telephone and the Farmer

The REA realized that, on a practical level, many rural Americans needed telephone service more than their city cousins because life on the farm often was isolated and far from doctors and other essential services. To drive this point home, the REA hired film director Irving Rusinow to make a movie dramatizing the need for telephone service in the country. After six months of shooting, the 28-minute color documentary, *The Telephone and the Farmer,* was released in December 1951.

The opening scenes of the film depict a farmer, played by Winnebago Rural Electric Cooperative Association employee Arthur Mitchell, who is hard at work shoveling snow so he can get his tractor out. "Then, according to the script, I fall down from a heart attack," Mitchell said, explaining that the emergency sets in motion the dire need for telephone service. "My movie wife has to dig out the tractor and then drive it into town."

In the film, the young farm wife reaches the doctor's office only to find he is not there. The next scene shifts to the general store, and an excerpt from the script depicts the citizens' frustration over the poor telephone service:

[Storekeeper answers telephone on counter] *"Hello! Hello, Roy. Are the roads open up there? I say, has the snowplow gone through yet? Dogonnit, Emma, can't you give us a better connection? Hello, Roy! Roy!"*

[Operator's voice (from other end of wire)] *"Why don't you just stick your head out the back window?"*

Mitchell laughed at the memory of that scene, but similar situations in real life were no laughing matter.

REA field agents showed the film hundreds of times throughout the country to inspire rural electric cooperatives to promote rural telephony. The REA came to shoot the film in Winnebago County, Iowa, at the urging of Glenn Bergland, then-manager of the Winnebago Cooperative Telephone Association, because Winnebago was an excellent example of forming a telephone cooperative. Bergland also promised the producers that Iowa could deliver the snow they needed for the script.

Although the film lived up to its promise to show the possibilities of rural telephony and even won an award at the International Edinburgh Film Festival, commercial telephone companies objected to the movie, arguing that it was biased toward cooperatives. The REA tried to appease these concerns by revising the film in 1953, but the film eventually was withdrawn from distribution.

Mitchell, now 84, said he is amazed at the telecommunications advancements he has seen in his lifetime. "I remember when getting dial telephones was a huge step. Now, you can take a cellular phone and talk to anyone anywhere in the world."

Strength in Numbers

s the REA telephone loan program got under way and more and more telephone cooperatives formed across the nation, the need for a national trade association to specifically represent REA telephone borrowers became apparent.

For several years, the National Rural Electric Cooperative Association (NRECA) had taken a leadership role on rural telephony issues. Beginning in the mid-1940s, the association passed resolutions at its annual meeting calling for its members to assist in the extension of modern telephone service throughout rural areas. In 1947, the NRECA formed a Communications Committee, which supported the concept of an REA telephone loan program. The association passed resolutions in 1948 and 1949 calling for such a program and took an active role in supporting the proposed legislation on Capitol Hill. Soon after the passage of the telephone amendments to the Rural Electrification Act, the NRECA Communications Committee became the NRECA Telephone Committee.

In 1952, in conjunction with an NRECA regional meeting in Lincoln, Nebraska, cooperative representatives elected state directors to serve as liaisons between the state and national levels on telephony issues. Then at the NRECA annual meeting in January 1954, definitive plans for a national organization began to materialize. Meeting in a hotel room in Miami Beach, Florida, interested individuals elected a committee to establish the organization and begin soliciting members.

On June 1, 1954, the group applied for and received a charter of incorporation under the District of Columbia Cooperative Association Act. The new organization—the National Telephone Cooperative Association (NTCA)—elected officers, adopted bylaws, and determined that the association headquarters would be at the NRECA building. The first national association for REA telephone borrowers had begun.

During a June 1954 organizational meeting of the newly formed National Telephone Cooperative Association, the early leaders adopt bylaws and elect officers.

When Disaster Strikes

Ice storms often resulted in disaster for REA borrowers, particularly in the days before buried plant became commonplace. Bledsoe Telephone Cooperative, Pikeville, Tennessee, suffered extensive damage during this storm in March 1960.

For most borrowers, receiving an REA loan and undertaking the construction of a new or expanded system brought a time of excitement and anticipation. But for some, it marked a time of great distress and anxiety.

The flood in Georgia, the earthquake in Alaska, the snowstorms in North Dakota, the tornado in Kansas ... the list of disasters goes on and on. Each time, the REA came to the assistance of borrowers whose systems and communities had been torn apart. The emergency loans for rebuilding were approved in record time, and the REA staff assisted in the restoration plans and construction. Realizing the importance of communications in the aftermath of a disaster, the REA also worked with the borrowers to establish temporary service as soon as possible.

One of the most detailed accounts of the REA's role in disaster recovery came in a speech by William Henning, president of Cameron Telephone Company, Sulphur, Louisiana, during a December 1957 industry meeting. Just months earlier, on June 27, Hurricane Audrey struck the Louisiana coast, bringing nearly a half million dollars in damage to the telephone system.

Henning told how the REA staff "worked day and night going in and out of the disaster area. ... before the authorities could start cleaning it up. They stumbled and stalked their way through the filth, dirt, and debris, trying to determine and appraise exactly what the situation was." After reviewing all the facts and figures, they decided to recommend a new loan for Cameron. "I truthfully believe had anyone else been our mortgagee, right after that storm we would have been sold for a percentage on the dollar," said Henning.

One week later, the loan was approved, and within three weeks, the Louisiana Public Service Commission had given its approval. The $480,000 in funds was available on July 22. Cameron used the loan to totally rebuild its system using primarily buried plant. Within 10 months, the company was back in service at a higher level than it had been prior to the storm.

"It is a tribute to REA to have the caliber and type of men who visited me and whose judgment and foresight, particularly under the most trying and difficult circumstances, is now proving to be correct," said Henning. "But at the time, REA went out on a limb, way out on a limb. ... REA was interested in our telephone company and it was interested in the re-establishment of service. ... This is the human side of REA. This is its personality; we may be a number to it for the file, but I never will believe that is all we are anymore. Whenever a crisis may arise, I think REA will be there to evaluate and see if it can possibly assist."

Overriding a Veto

Despite the telephone program's obvious success in serving its borrowers, the REA faced some rather strong opposition from the Eisenhower administration. Although a letter from the President to REA Administrator Ancher Nelsen in late 1954 outlined the administration's support for the program, subsequent actions showed otherwise. In 1957, the administration proposed legislation to more than double the REA 2 percent interest rate, and the following year, a measure for private-market financing. Fortunately for the agency and its borrowers, the first bill did not go anywhere and the second was not even introduced.

But the administration dealt an even bigger blow in 1957 when Secretary of Agriculture Ezra Benson shifted loan-making authority away from the REA administrator, requiring the top Agricultural Credit Services staff to review all loans in excess of $500,000. The action ran directly contrary to Benson's pledge not to interfere with the REA's loan-making authority, which he made when Congress approved a reorganization of the Department of Agriculture in 1953.

In an effort to correct the situation, Representative Melvin Price (D-IL) and Senator Hubert Humphrey (D-MN) introduced legislation calling for loan-making authority to be restored to the REA. While the bills passed both the House and the Senate in early 1959, the President vetoed the Humphrey-Price measure on April 27, 1959.

Grass-roots efforts by both telephone and electric borrowers went into full swing, and with a great measure of success: the next day, the Senate overrode the presidential veto 64-29, and a few days later, the House came within four votes of doing the same. The political pressure proved enough to force the administration to curtail its anti-REA efforts.

Senator Hubert Humphrey (D-MN) stood as one of the REA program's strongest supporters, not only during the crisis of the late 1950s, but throughout his career. After his address during the National Telephone Cooperative Association (NTCA) 14th Annual Meeting in 1968, then-Vice President Humphrey greets NTCA Director Robert Jamison of Horry Telephone Cooperative, Conway, South Carolina.

As part of the REA telephone program's week-long 10th anniversary celebration in 1959, representatives of the REA and the telephone industry show they are "united for progress." Pictured (from left) are REA Deputy Administrator Ralph Foreman; Oral Moody, AT&T; REA Administrator David Hamil; William Henry, president and general manager of Northern Ohio Telephone Company, Bellevue, Ohio; and David McKay, manager of Golden Belt Telephone Association, LaCrosse, Kansas.

Making Some Changes

During the early days of the REA telephone loan program, it operated as a branch of the electric program, and staff members worked on both telephone and electric issues. But realizing the significant differences between the two utilities, the REA reorganized in mid-1952, establishing a separate division to concentrate on the telephone loan program. The agency increased the telephone program's staff by 50 percent. Then in 1953, the REA established the position of deputy administrator, as well as two assistant administrator positions—one for telephone and one for electric.

As the decade moved on, the REA realized some of its earlier decisions had not been the best ones. In terms of loan security, the agency decided developing sound management and operations meant more than high levels of equity. Thus in 1957, at a meeting of the National Telephone Cooperative Association, REA Administrator David Hamil announced that the $25 to $50 equity requirement for telephone cooperatives would be reduced to $10. The new requirements called for cooperatives to have sign-ups equal to 70 percent of the five-year subscriber estimate and collect $10 from each unserved applicant, while commercial companies needed to provide 10 percent equity. This move helped address the growing industry concern about the REA telephone program favoring commercial companies.

The agency made several other adjustments to its telephone loan program structure during the 1950s, including lessening its oversight of borrowers' operations and streamlining the overall loan process.

Meanwhile, the REA had begun to face instances of unflattering media coverage, both in the trade and general press. In 1951, popular magazine *Reader's Digest* carried a brief article criticizing the REA loan program for approving a loan for Farmers Telephone Company, Iowa, in an amount greater than what the telephone company had requested. The higher loan amount derived from the agency's requirement that borrowers have no more than eight parties per line.

Critics accused the REA of "gold-plating" rural telephone facilities; they claimed rural people did not want or need such "fancy" service. Congress did hold hearings on the affair, but when the REA explained that eight-party service was necessary to permit full selective ringing, Congress put the matter to rest.

The REA again made the news later that year, but this time in the wake of a tragedy. On June 30, 1951, six REA staff members, traveling on REA business, were killed in a plane crash. Among them was Deputy Administrator George Haggard.

Williston Mayor Quincy Kennedy stands solemnly by as Byron Wham, owner of Williston Telephone Company, Williston, South Carolina, sadly mourns "an old handcrank telephone" during a 1952 cutover celebration.

A Cause to Celebrate

As the number of REA telephone loans grew, the cutovers of exchanges to dial service became more and more frequent. Cutover day meant a time for celebration—and in many cases, a time for demonstration and instruction, because the new dial telephone instruments were foreign to most subscribers. Throughout the nation, telephone systems took a variety of creative approaches to this special occasion. Many shared their experiences with the REA, who in turn reported them in its *Rural Electrification News* magazine for other systems to consider.

One common ceremony was the mock funeral, where attendees, often dressed in black, "mourned" the passing of their old crank phones. The funeral generally included a dug grave and even a tombstone with a fitting epitaph.

Also popular were demonstrations in the high school auditorium, as this provided ample seating and a stage on which to carry out the day's events. These programs generally included a demonstration of how to use the telephone, in addition to the traditional "first call."

Eureka Telephone Company in Corydon, Indiana, took a more personal approach to its cutover. The company invited subscribers to a series of teas throughout the community, where each individual had a chance to try using the telephone.

Borrowers also shared with the REA the stories of their customers' reactions to the new dial telephone service. Most told of excited customers who made many more calls than they had made on their old magneto sets and who extolled the numerous ways in which the instrument improved their work and home lives. Eastex Telephone Cooperative, Laneville, Texas, for example, quoted one customer: "We pay the bill with the gas money we save!"

Some subscribers were, understandably, a bit nervous and confused by the new telephone instruments, and still other subscribers expressed sadness about the change. They missed the informality of their former service, especially hearing the voice of the familiar operator, and in the case of single-party service, they missed the ability to listen in on one another's conversations!

William Henning, president of Cameron Telephone Company, Sulphur, Louisiana, watches as Mrs. Bonsall makes the first dial call to Cincinnati, Ohio, during a cutover ceremony held in the Cameron High School auditorium in 1954.

Operator Mrs. L.E. Bass connects a call at the switchboard in the McAdoo, Texas, exchange of the South Plains Telephone Cooperative, Lubbock, Texas.

Operator Riola Knight uses her ringing key to ring a subscriber from the switchboard at Gray Court Telephone Company, now part of Piedmont Rural Telephone Cooperative, Laurens, South Carolina.

The End of an Era

Saying hello to a new dial telephone system meant saying goodbye to a faithful part of a rural telephone system—the switchboard operator. Many newspapers covering an REA borrower's pending upgrade to improved service also carried an article featuring the person who had worked the switchboard year after year. Most operators were women; while boys initially held such positions, their bent to swearing and roughhousing soon converted the field to an employment opportunity for women.

It was no wonder people missed the operator, as she did so much more than connect telephone calls. In most communities, she was a message taker, locator service, and news reporter, in addition to performing countless other duties.

Many REA publications and telephone systems' historic documents included accounts of the local operator's work: ringing everyone's line at 7 p.m. to report the correct time; having the local weather forecast, sporting event results, local and national news, and even market quotations available for those who asked; announcing advertisements and store specials; and being able to answer questions about community matters—the location of fires, the time of church meetings, who had died, and the names of the new people in town. She often was responsible for spreading the alarm when a fire broke out and had the unpleasant job of relaying death notifications during the war.

The operator generally knew where people were day or night, and it was not uncommon for people to tell the operator where they were going and for how long, or to call to the operator to determine someone's whereabouts. Sometimes this even meant her physically looking out the window to answer questions such as "Have you seen my husband up town?"

Because of such familiarity, most people did not ask to be connected to a specific number when placing a call, but rather asked for the person by name or even simply "Grandpa" or "the general store." The operator memorized the telephone directory and learned all family members' names.

Some requests were a bit more specialized in nature: "Ring me up in 20 minutes so I won't forget the pie," "I put the receiver in cradle. Ring me up if the baby wakes up and cries," and "Ring me a half hour before the 4:20 train and first check to see if it's on time."

Often the operator lived in the building that housed the switchboard. She had designated relief people, but worked long hours—generally day and night—for modest wages.

With the advent of the REA telephone program and the coming of dial service, the switchboards found their way to storage rooms, junkyards, and antique shops, while the familiar voices answering "Central …" and "Number please …" faded away forever.

Facts and Figures: *REA/RUS Standards and Engineering*

From the moment the REA telephone program came into existence, its engineering staff faced the formidable challenge of developing rural solutions, of filling the obvious void of standards, engineering methods, and construction techniques specifically for rural telephone systems. With its countless accomplishments, the agency not only met but exceeded the challenge. Many believe the agency's biggest contribution to be its standards and specifications for rural equipment design and construction techniques, even more so than the low-cost capital it provided. The following list, while far from complete, summarizes several REA standards and engineering accomplishments and innovations through the years.

Assembly Units. Early on, REA engineers developed construction assembly units for all telephone plant items, allowing contractors to bid competitively on construction.

Dial Switching Equipment. In fiscal year 1951, the REA developed a standard technical specification for automatic dial switching equipment—the first such standard in the industry. This allowed borrowers to obtain competitive bids from central office equipment manufacturers and ensured the lowest possible cost.

Carrier Equipment. From the start, the REA took the lead in the development of subscriber carrier systems, and through the years, the agency worked with borrowers and manufacturers to continue to increase the technology's potential. Basically, carrier equipment uses electronics to increase the number of conversation paths that can be carried on a single pair of wires. This technology allowed borrowers to serve sparsely populated areas long distances from their central offices cost effectively.

Microwave Radio. The REA pioneered work in the use of microwave for getting telephone service to hard-to-reach places. It worked to develop a low-cost, point-to-point microwave radio system because existing microwave technology proved too costly for small installations.

Selective Ringing. After testing various types of ringers and gathering data, the REA in 1953 contracted for the development and construction of equipment for fully selective ringing for rural party-line service. After some modifications to the approach, frequency selective ringers were developed.

In November 1959, the REA conducted a field trial of its long-span construction theory, using insulated open wire, at Tyler Telephone Company, Tyler, Minnesota. The concept called for making span lengths long enough and the wire just high enough above the ground so that when heavy ice formed on the line, the wire would rest on the ground. During this storm, the span had an ice load one inch in diameter but suffered little damage, and service remained intact, even though much of the wire was buried under the snow. To keep lines in service during normal usage hours, much testing took place between 11 p.m. and 4 a.m., outdoors in blizzard conditions and subzero temperatures.

Telephone Sets. Because REA borrowers could not buy equipment from Western Electric, they had to rely on independent telephone set manufacturers. And because no standards existed, the REA tested the sets to ensure they indeed performed as well as the Bell System equipment. To evaluate the performance of a wide variety of sets, the REA established an acoustical testing lab in the Cotton Annex, located across the street from the U.S. Department of Agriculture South Building. The test results allowed the agency to eliminate poorly designed equipment and thus be sure its borrowers used only the best telephone sets in their systems. Also, the U.S. Signal Corps tested telephone sets on behalf of the REA under uniform procedures and conditions. The REA continued to test the new types of sets being manufactured through the years to ensure their compatibility with changing system technology.

Telephone Poles. In 1953, the REA engineers developed and field tested a pole-top assembly using an 18-inch, two-pin crossarm instead of the traditional wooden bracket construction. It allowed for simpler mounting, could be placed higher on the pole, and could easily be moved to a new location. It also permitted greater wire spacing, which helped reduce power line interference.

Mobile Radio. Mobile service in the 1950s required an operator to complete the call, so when REA-financed telephone systems converted to automated dial offices, mobile service would not work—there was no operator. Thus, the REA in 1956 contracted for the development of mobile radio equipment that would permit direct dialing. Automatic Electric developed equipment, later marketed by Lenkurt, that used a rotary dial, while Motorola's equipment used pushbutton dialing (but not the same as touchtone). Both systems, installed in 1957, performed well but required some modifications. Soon service was in use throughout the nation, and over time it evolved into improved mobile telephone service (IMTS). The REA's pioneering work in mobile dial telephone service in the late 1950s is a key factor in mobile telephone service being what it is today.

Plastic Insulated Cable. In 1954, the REA developed a specification for plastic insulated cable, the first such specification available for this cable and thus used throughout the telephone industry. REA-generated research and development also resulted in a splicing connector and other accessories for plastic insulated cable.

Buried Plant. The 1950s brought the introduction of buried plant, and REA engineers worked hard to perfect the design and application of standards for its use in rural systems. With buried plant's lower initial costs and significantly reduced maintenance costs, it soon made open wire construction obsolete. The agency continued to improve techniques for buried plant and accomplished a great deal in terms of developing and setting standards for buried plant products and installation techniques.

Cable Loading. As system design shifted from wire to cable construction in the late 1950s, there arose a need to minimize voice transmission losses. REA engineers developed the D-66 cable loading system, which became a

North Florida Telephone Company, Live Oak, Florida, and Central Virginia Telephone Corporation, Amherst, Virginia, conducted the field trials of the first dial mobile radio equipment built to REA specifications. North Florida tested the rotary dial equipment, while Central Virginia tested the pushbutton version. Both obtained waivers of the Federal Communications Commission rules to allow operation without the use of an operator and call signs. Here, Central Virginia President and Manager L. John Denney uses his mobile unit to place a call through the Raphine, Virginia, exchange. The telephone systems also used the radio equipment to provide fixed service in remote locations.

standard for rural telephone transmission. Because the frequency range of the D-66 system more closely matched the range for intelligible speech, conversations traveling over long rural loops actually proved more understandable than those carried over shorter Bell company loops.

Long-Span Construction. In 1959, the REA developed long-span construction—using higher tensile strength wire and half the number of poles. The lines stretched instead of breaking during heavy snow and ice storms, and the reduced number of poles cut construction costs. The cable included a polyethylene coating to address the problems of mid-span contact between the wires and of contact with the ground when it stretched.

Color Coding. In the late 1950s, the REA developed a uniform color coding system for identifying wire pairs in cables. It became an REA standard and was widely adopted throughout the communications industry.

Fine-Gauge Design. In 1963, the REA introduced its fine-gauge design, which allowed systems to use 24-gauge cable even over long distances where 19- or 16-gauge cable previously had been required. Electronic devices, such as long-line adapters and loop extenders, increased the signaling range and transmission efficiency of the thinner lines, and low-cost voice frequency repeaters amplified the signal, all making the new design possible. The thinner cable contained much less copper, which had become increasingly scarce and expensive.

Common Mode Operation. The REA helped initiate this new telephone design technique in 1966. It placed the repeater and adapter inside the central office switching equipment, thus making it possible to share them on a common basis with other subscriber loops served by the same office. This not only reduced the number of repeaters and adapters needed, but also allowed for greater use of fine-gauge cable.

Lightning Testing. Because rural telephone systems are more exposed to lightning damage, the REA believed it important that equipment be capable of withstanding lightning surges. REA engineers developed requirements for various types of equipment and actually designed and built a generator to simulate lightning strikes for testing, which the agency housed in the Cotton Annex.

Filled Cables. A real plus came to the outside plant arena in the late 1960s and early 1970s with the introduction of grease-filled cables. Filling the air spaces with a petroleum substance alleviated the moisture problems inherent in underground construction and thus lowered maintenance costs substantially. The REA played a key role in this development, and filled buried cable became the REA standard in 1974.

Digital Switching Equipment. The REA's digital switch specification, issued in 1978, was the first in the industry. It addressed the problem of digital central offices only connecting digitally with other equipment made by the same manufacturer, thus opening the way for borrowers to purchase system components from multiple vendors.

Workers at East Ascension Telephone Company, Gonzales, Louisiana, dig holes for anchoring guy wires, the wires that connect to the tops of telephone poles to hold them upright.

In the 1960s, the REA took on a renewed sense of purpose as its focus shifted to upgrading existing service and striving for rural service comparable to urban service. The technical innovations continued, and as a result, the REA raised its minimum standards for system design. The agency also began to emphasize borrower development, with the goal of nurturing a level of strength and self-sufficiency that would ensure a permanent existence. And the agency enjoyed perhaps its strongest presidential support ever under the Johnson administration. But soon the backlog of loan applications began to mount, underscoring the need for supplemental financing for the agency to continue its goals of …

Parity, Progress, and Permanence

Just One Party

T he REA continued its work in the areas of technology development and standards setting, allowing borrowers to deploy increasingly sophisticated, yet cost-effective, telephone systems. Such was its success that in mid-1963, the agency announced it had upgraded its minimum standard to four-party service or better, instead of eight-party. The agency additionally urged borrowers to prepare to meet the demand for one- and two-party service as it developed.

Just two years later, REA Administrator Norman Clapp issued a press release stating that the agency strongly encouraged the installation of single-party service. He noted that the move to buried plant, combined with other construction improvements, had reduced the cost of building a circuit mile of line from an average of $217 in 1959 to $98 in 1964, with an anticipated drop to only $70 by 1970.

A key player in the shift to single-party service was REA borrower Dunnell Telephone Company, Dunnell, Minnesota. When applying for a loan to upgrade its 300-phone system in 1963, the company and its engineers submitted an area coverage design that contained two plans—one for two- and four-party service, and the other for all one-party service.

The REA's loan feasibility studies revealed that the cost differential was not significant, so the agency said it would finance an all-one-party system if the Minnesota Public Service Commission (PSC) approved the plan as well. A PSC survey of subscribers found that 95 percent of them preferred one-party service. Thus, the state commission gave the nod, and Dunnell became the first REA borrower and the first rural telephone system in the nation to provide all one-party, buried rural service. The REA approved the $150,000 loan on October 31, 1963, and Dunnell cut over its system the next fall.

The REA soon learned that with fewer parties on the line, subscribers made more long distance calls, which meant additional revenue to offset the cost of one-party service. The agency also witnessed fewer trouble reports, reduced maintenance, and an increased use of extension phones and other services.

A Few More Adjustments

By the early 1960s, the industry had begun to raise some concerns about the operation of the REA telephone loan program, particularly in terms of what many perceived as a bias against cooperatives. They pointed to the diminishing number of loans allocated to cooperatives and claimed the REA enforced its equity requirements much more strictly for cooperatives than for commercial systems.

When Norman Clapp became REA administrator in 1961, he announced plans to reappraise the entire telephone program and look closely at these issues, as well as address the agency's own concern about the speculative activity that had begun to emerge in the program. People were using REA funds to make a quick profit rather than to provide area coverage service.

The REA in 1962 implemented a new policy calling for local ownership and control of telephone companies. The policy stated that if a company were not locally owned, it must meet a minimum level of net worth, which meant a higher equity payment. Also, any acquisitions had to be logical extensions of the telephone system's operations.

These requirements, designed to encourage local ownership and the formation of cooperatives, did help get the cooperative portion of the telephone program moving again, as the percentage of loans going to cooperatives increased during each of the next several years.

A few years later, the U.S. Treasury made an REA series of 2 percent treasury bonds available only to REA borrowers. The bonds matured in 12 years, but could be cashed on 30 days' notice. This gave borrowers the opportunity to lend to the government at the same interest rate they received on their loans and was aimed at halting the criticism that had begun to grow concerning borrowers' available cash and their investments.

Rural Electrification Act amendments adopted in 1962 allowed the REA to lend for non-voice communications, including educational television and related equipment installations. Shortly thereafter, the REA approved three loans to finance technical equipment to relay educational television programs to schools. St. Matthews Telephone Company, St. Matthews, South Carolina, was the first to provide closed circuit educational television in March 1963. Using coaxial cable, the system connected four high schools in the area. Programming originated 37 miles away in Columbia, South Carolina, and brought advanced classes to 800 students, including this group of math students.

Building Better Borrowers

A significant effort in the 1960s focused on the REA guiding policy stating that the agency "shall render certain technical advice and assistance to its borrowers" and that this help "shall progressively diminish as borrowers gain in strength and maturity." The REA's goal was to move to a point at which "every borrower possesses the internal strength and soundness to guarantee its permanent success as an independent local enterprise." To solidify its commitment, the REA in 1966 established an assistant administrator for borrower development who based his work on the agency's realization that every borrower was different.

To implement this guiding policy, the REA continued and increased its multifaceted technical assistance program, as well as its management and operational assistance. The agency also stepped up its focus on customer service, with an eye toward increasing borrower revenues. In 1965, the REA extended its Five-Star Member Service Program, previously developed for electric cooperatives, to telephone cooperatives. Implemented with the assistance of REA field representatives, the program helped assure that all cooperative members received the privilege of ownership; the benefits of non-profit operations; adequate, dependable, low-cost telephone service; advice on the full use of telephone service; and information concerning the cooperative and its operation.

The REA also began to encourage borrowers to jointly implement new office technologies such as automatic data processing equipment.

The success of the REA telephone loan program did not stay a secret. Telecommunications officials from many foreign countries came to visit the REA offices, as well as borrower facilities, for consultation and training to learn how they could emulate the REA program abroad. The agency also responded to numerous written inquiries, primarily on engineering and standardization issues. And REA staff specialists—both electric and telephone—traveled throughout the world to share their expertise, often through the Agency for International Development. The REA pattern had become an exportable item, and the REA began to play a role in the development of telephone systems in other countries.

On the occasion of the REA telephone program's 20th anniversary in 1969, the industry presented a rare magneto telephone mounted on a wall plaque to the House Committee on Agriculture. Committee Chairman W.R. "Bob" Poage (D-TX) (center) accepts the telephone from National REA Telephone Association Executive Director A. Harold Peterson (left) and National Telephone Cooperative Association (NTCA) Executive Manager David Fullarton. NTCA Director James Wright, Cap Rock Telephone Cooperative, Spur, Texas, donated the telephone.

Equal Opportunity Employer

Dr. Henry Taylor, the REA's first civil rights coordinator, developed a variety of programs to assist REA borrowers in increasing minority participation on their boards of directors, as well as within their company operations.

An Opportunity for Everyone

I n addition to its tireless work in the areas of telephone engineering, construction, and operations, the REA also took action when it came to affirmative action. Over the years, the REA/RUS has worked to ensure minorities receive loans, telephone service, and employment opportunities.

As early as 1948, the REA loan contract included a statement prohibiting borrowers from discriminating against any employee due to race, creed, color, sex, or national origin. Every vendor contract involving employment negotiated by borrowers contains a similar statement.

Beginning in the 1960s, the REA issued a series of policy bulletins on non-discrimination and equal opportunity employment, urging and re-emphasizing the responsibility of every REA borrower to operate without regard to the race, creed, color, sex, or national origin of its employees and subscribers.

To drive this point home, the REA ruled in the early 1970s that borrowers with 50 or more employees must develop and carry out a written affirmative action program. The agency also instituted a program of civil rights reviews, including borrower reporting forms and REA field visits to several borrowers each year. These review activities continue to this day.

The REA's first civil rights coordinator, Dr. Henry Taylor, worked hard to establish guidelines and suggestions to help borrowers meet the requirements of the equal opportunity statutes in the Civil Rights Act and subsequent executive orders. Taylor believed that "the equitable involvement of minority groups and women, and the overall employment requirements of rural electric and telephone systems, should not be left to chance."

Taylor recommended to systems that they take a positive approach and outlined simple steps they could take to increase minority participation in their operations. His many articles and communications to borrowers stressed the good business aspects of civil rights activities rather than their being required by law. By the mid-1970s, borrowers had begun to exhibit progress, as the number of minorities on boards of directors continued to increase.

The RUS continues to encourage affirmative action plans among borrowers, emphasizing that if a large percentage of their subscribers are minorities, they should have minorities on their boards and employed in their offices. The agency has received very few complaints about telephone companies in the areas of employment or service.

The Rural Development Role

The REA telephone loan program in and of itself stood as a prime example of rural economic development, bringing jobs to rural areas, increasing the tax base, and providing communications services that made the areas more attractive to new residents and businesses. But the agency also realized that any additional steps borrowers could take to improve the economy of their service areas would improve their financial positions—and the REA's loan security. Thus, from the beginning, the REA emphasized the need for rural development efforts.

With the creation of the Rural Areas Development (RAD) program within the Department of Agriculture and the passage of the Area Redevelopment Administration (ARA) legislation, both in 1961, the REA's rural development role greatly expanded.

As a RAD agency, the REA assisted its borrowers with community development activities. The REA told borrowers how to help new businesses find facilities and financing sources; how to offer their services, meeting facilities, and employee skills for community projects and civic groups; and how to lead area development efforts.

To better accomplish its goals, the REA established a RAD staff in the office of the administrator and expanded its Section 5 loan program, which had been established for electric borrowers to lend money to consumers for wiring and appliance purchases. But this last move brought some unfortunate results. The agency hoped to give rural development efforts a boost by allowing borrowers to secure funds and then relend them to communities for financing the equipment needed for RAD projects. But loans to ski resorts for snowmaking equipment, while they accomplished exactly what the REA intended on the rural development front, became the focus of much negative press, even congressional hearings. The agency was forced to curtail its use of the Section 5 loan program for such activities.

The ARA measure established loans and grants to assist designated rural counties, 95 percent of which were served by REA electric and/or telephone borrowers. The REA became an ARA delegate agency assigned responsibility for carrying out certain rural provisions of the new law, including the review of project proposals.

The REA surveyed borrowers about their rural development activities, and according to the fiscal year 1964 *Report of the Administrator*, REA electric and telephone borrowers had helped launch more than 900 industrial and business enterprises, creating more than 60,000 direct jobs and more than 40,000 indirect jobs, from July 1961 to the end of 1963.

Secretary of Agriculture Orville Freeman and members of the Cooperative Advisory Committee—the U.S. Department of Agriculture's (USDA) public consultants on its work with cooperatives—tighten the "Knot of Cooperation" to signify the role of the USDA and cooperatives in helping rural people to build a strong rural America. The group participated in the opening ceremonies of the USDA's Co-op Month Observance, October 6-23, 1964. The pullers include (from left) Assistant Secretary of Agriculture George Mohran; Under Secretary of Agriculture Charles Murphy; Secretary of the Treasury Douglas Dillon; Freeman; Clyde Ellis, National Rural Electric Cooperative Association; J.K. Stern, American Institute of Cooperation; Representative Graham Purcell Jr. (D-TX); Kenneth Naden, National Council of Farmer Cooperatives; Dwight Townsend, The Cooperative League of the USA; and Patrick Healy, National Milk Producers Federation. Also on the rope, but not shown, are Assistant Secretary of Agriculture John Baker and Roy Hendrickson, National Federation of Grain Cooperatives.

Making the Five Million Mark

Congratulations! You're our five millionth customer! The July 2, 1962, celebration marking the connection of REA electric subscriber number five million carried all the hoopla and excitement of a modern-day sweepstakes award. On that day, the media and national, state, and local dignitaries, as well as family and friends, descended upon the ranch home of Mr. and Mrs. John McGuffin, located 30 miles outside Tatum, New Mexico.

The celebration also brought some notoriety to the REA telephone program, for the McGuffins also received telephone service for the first time that day. The Lea County Electric Cooperative and the Leaco Rural Telephone Cooperative used the same poles to extend their services 3.5 miles to the ranch at a cost of $4,000. There was, however, no additional charge to the subscriber due to the REA's area coverage policy.

REA Administrator Norman Clapp was on hand to actually "turn the switch" for the McGuffins' power, and R.B. Moore, manager of both cooperatives, presented the couple with several appliances. Secretary of Agriculture Orville Freeman placed a call to the McGuffin ranch during the celebration, and everyone in attendance enjoyed a hearty chuck wagon meal.

Not Enough Money

B y the mid-1960s, the need for capital had begun to outstrip the REA telephone loan program's supply. At the end of fiscal year 1965, the loan applications remaining on hand totaled $25 million more than the amount loaned during the year, and by a year later, the pending applications amounted to double the year's loan program. The high demand, coupled with the general economic condition of the nation due to the Vietnam War, made funding tight.

To ensure funds for as many projects as possible, the REA began a program of selectivity, analyzing the specific needs of each borrower and trying to trim loan requests to just the amount necessary. In 1962, the agency adopted a policy requiring borrowers with general funds exceeding 20 percent of their plant to make accelerated loan payments "to prevent excessive accumulation of general funds by borrowers."

By 1966, the agency had officially implemented its Joint REA-Borrower Cash Management Program. The REA asked borrowers to defer any construction that could be postponed, make maximum use of their general funds, and continue advance payments, all in an effort to reduce the cash flow out of the U.S. Treasury and combat inflation. While the program proved quite successful, a large level of unmet loan demand remained.

Soon it became clear the problem required a legislative approach. Anticipating an inevitable change to the program, the National Rural Electric Cooperative Association contracted investment firm Kuhn, Loeb, Inc. in 1964 to study the various possibilities for supplemental financing for electric cooperatives. The telephone industry used the same firm for a study the following year, and the administration began developing its legislative proposal.

The legislative effort began in 1966 with the introduction by Representative W.R. "Bob" Poage (D-TX) of a measure to create two federal banks for supplemental financing—one for electric borrowers and one for telephone borrowers—but also retain the 2 percent loan program. The administration sent a similar measure to Capitol Hill shortly thereafter.

Both the 1966 legislation and similar measures introduced in 1967 failed, the first due to lack of time, and the latter due to electric industry concerns. Thus, later in 1967 and again in 1969, Poage introduced a bill covering only the telephone program. While both times the bill died waiting for the House Rules Committee to schedule it for floor debate, many ideas contained in it later were incorporated into the measure creating the Rural Telephone Bank in 1971.

National Telephone Cooperative Association (NTCA) President B. Maynard Christenson, Dakota Cooperative Telephone Company, Irene, South Dakota (left), and NTCA Executive Manager David Fullarton appear before the House Agriculture Committee on March 23, 1967, to testify on the need for supplemental REA financing.

Facts and Figures: *REA/RUS Officials*

The following are the individuals appointed to the top staff positions for the REA/RUS telephone loan program since its inception. (Please note: This list does not include the many individuals who filled these positions in an acting capacity pending the appointment of several of the listed individuals.)

Administrators

Claude R. Wickard
1945-1953

Ancher Nelsen
1953-1956

David A. Hamil
1956-1961

Norman M. Clapp
1961-1969

David A. Hamil
1969-1978

Robert W. Feragen
1978-1981

Harold V. Hunter
1981-1988

Gary Byrne
1990-1991

James B. Huff
1992-1993

Wally Beyer
1993-present

Deputy Administrators

George W. Haggard
1949-1951

William C. Wise
1951-1953

Fred Strong
1954-1958

Ralph Foreman
1958-1961

Deputy Administrators (cont.)

Richard A. Dell
1961-1966

Richard M. Hausler
1966-1969

Everett C. Weitzell
1969-1973

George P. Herzog
1973-1974

David H. Askegaard
1975-1976

Robert W. Feragen
1977-1978

Susan T. Shepherd
1978-1980

Jack Van Mark
1981-1988

George E. Pratt
1989-1993

Adam M. Golodner
1994-1997

Christopher A. McLean
1998-present

Assistant Administrators - Telephone*

J.K. O'Shaughnessy
1953-1957

Norman H. McFarlin
1958-1960

Frank Renshaw
1961-1973

Charles R. Ballard
1973-1977

John H. Arnesen
1977-1987

Charles R. Miller
1987-1988

Robert Peters
1989-1997

Roberta D. Purcell
1998-present

*William S. Rummens
succeeded O'Shaughnessy
in 1958 but only held the
position for a few weeks.

Deputy Assistant Administrators - Telephone

Walter L. Wolff: 1961-1971**
Raymond W. Lynn: 1969-1972**
John H. Arnesen: 1972-1978
William W. Kelly: 1978-1982
Anne K. Hanlon: 1982-1986***
William F. Albrecht: 1985-1989***

Barbara L. Eddy: 1989-1997
Jonathan P. Claffey: 1998-present

** During 1969-1971, Wolff was deputy for operations,
   while Lynn was deputy for engineering.
*** During 1985-1986, Hanlon and Albrecht held the
   position together.

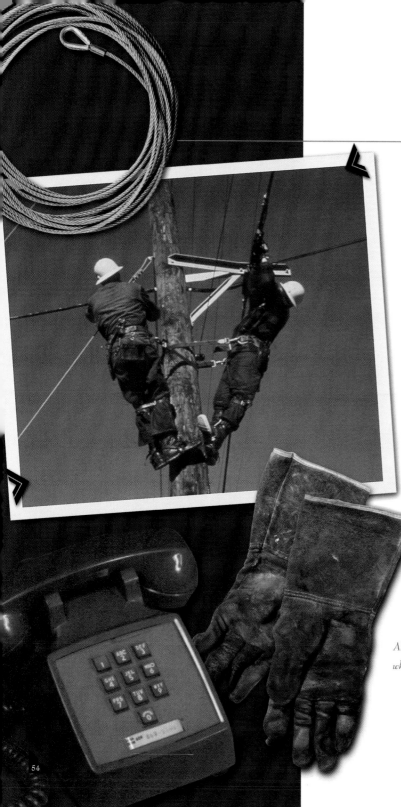

I nflation, consumerism, the energy crisis. During the 1970s, these general trends combined with several monumental changes and additions to the REA telephone loan program, making it a decade the REA and its borrowers will never forget. But despite some difficulties, the agency recorded a record number of "firsts" and presented many exciting new opportunities to its borrowers. And by late in the decade, the industry had entered a phase of rapidly changing technology, prompting borrowers to initiate new services instead of just reacting to customer demand. From start to finish, the 1970s brought …

Challenges and Changes

Although buried plant had become the standard, there remained times when aerial construction still proved the most appropriate.

Supplemental Success: The RTB

A s the decade got under way, the REA loan application backlog remained extremely heavy—by 1971, it was more than four times the average annual REA appropriation. Thus, in addition to its existing loan evaluation guidelines, the REA continued to ask borrowers to trim their loan requests to only the amount truly needed and to use their own money when possible. Meanwhile, legislative efforts to resolve the funding situation continued on Capitol Hill.

After years of roadblocks from the House Rules Committee, real progress on supplemental financing legislation began to materialize early in 1971. After some strong lobbying and political maneuvering by telephone borrowers and their trade associations prevented yet another halt by the House Rules Committee, both the House and Senate passed the measure, and a conference committee ironed out several differences.

Then, on May 7, 1971, President Richard Nixon signed Public Law 92-12, adding Title IV to the Rural Electrification Act to establish the Rural Telephone Bank (RTB). The law set up the RTB as a U.S. agency managed by a governor—the REA administrator—and a 13-member board of directors. In addition to the governor, the board included four other government representatives, two members from the general public appointed by the president, and six telephone industry representatives—three from commercial telephone companies and three from telephone cooperatives. The RTB was to use the staff and facilities of the REA and other Department of Agriculture agencies.

The legislation established a 10-year capitalization program for the bank, with the government investing $30 million each year. The measure also established three classes of RTB stock: Class A, which is owned by the federal government; Class B, which all RTB loan recipients are required to purchase (5 percent for every dollar borrowed); and Class C, which is available for purchase by entities eligible to borrow from the RTB. Also, beginning in June 1972, the bank gained the authority to obtain funds from the sale of its debentures to the U.S. Treasury.

While the bank is owned by both the government and its borrowers, the law established eventual borrower ownership and control through a privatization process of the government retiring its Class A stock, which began October 1, 1995, as set forth in the law.

Initially, the interest rate for RTB loans ranged from 4 to 8 percent, based on the borrower's ability to pay. And if a borrower could afford more than the REA direct loan rate of 2 percent, but not the RTB rate, the borrower received a combination of REA and RTB loans. REA Administrator David Hamil signed the first five RTB loans—totaling nearly $5 million—on January 21, 1972. Recipients included Eastex Telephone Cooperative, Henderson, Texas; Mutual Telephone Company, Sioux Center, Iowa; Pioneer Telephone Association, Ulysses, Kansas; Plant Telephone and Power Company, Tifton, Georgia; and Randolph Telephone Company, Liberty, North Carolina.

During a small ceremony in the Cabinet Room of the White House after the signing of the Rural Telephone Bank Bill, National Telephone Cooperative Association (NTCA) representatives present President Richard Nixon with a 1902 Kellogg antique magneto telephone. Those in attendance include (from left) Representative Ancher Nelsen (R-MN); NTCA President Eldon Snowden, McDonough Telephone Cooperative, Colchester, Illinois, who donated the phone from his personal collection; REA Administrator David Hamil; NTCA Executive Vice President David Fullarton; Secretary of Agriculture Clifford Hardin; Representative W.R. "Bob" Poage (D-TX); Representative Robert Price (R-WI); Senator Robert Dole (R-KS); Nixon; Senator Jack Miller (R-IA); and Representative Vernon Thomson (R-WI). Hidden from view is REA Deputy Administrator Everett Weitzell.

And the Winner Is ...

COOPERATIVE BOARD POS...

BALLOTING NO.→	1	2	3
COOPERATIVE BALLOTS			
NO. NEEDED FOR ELECTION	729		
	365		
1. BERGLAND, GLENN W.	(475)		
2. BOECHER, ROY C.	(373)		
3. CHRISTENSON, B. MAYNARD	(554)		
4. HOLLAND, R. MAURICE	348		
5. MCKAY, DAVID J.	303		

VOTE

*O*n September 15, 1971, President Richard Nixon appointed the original six industry representatives to the Rural Telephone Bank (RTB) board of directors. A year later, on September 14, 1972, the industry participated in the first RTB board election.

This election was unique in that all telephone entities eligible to borrow from the RTB could vote, and each could vote for all six industry seats. In all subsequent biennial elections, only actual RTB stockholders—borrowers and Class C stockholders—could participate; only one vote was allowed per holding company; and commercial and cooperative borrowers could vote only for the seats in their respective segment of the industry. The special rules for the first election proved unpopular with many in the industry.

The first election, held in Washington, D.C., developed into quite an industry event. Representatives from 146 systems registered to vote on election day and qualified to cast a total of 733 votes, including proxies. This represented 85 percent of the eligible voters. Six candidates ran for the three cooperative seats, while 10 vied for the three commercial seats.

The Rural Electrification Act called for board members to be elected by a majority, not a plurality, which led to the potential for several rounds of voting. In the 1972 election, all three cooperative members were elected on the first ballot, but only one of the commercial members garnered sufficient votes. The remaining two were not elected until the fourth ballot.

Two years later, on September 19, 1974, the first of the bank elections under the standard rules took place in the nation's capital. Gone was the excitement of the first election. About 260 of the just more than 300 stockholders participated in the election in person or via proxy, and the election was completed after the first ballot.

The vote tally board for the race for the three cooperative positions in the first RTB election clearly shows the three winners after just one ballot.

The Event at Squaw Gap

When Reservation Telephone Cooperative, Parshall, North Dakota, cut over its Squaw Gap exchange in 1971, it not only brought one-party dial service to 100 citizens who never before had a telephone, but also brought national attention to the REA telephone loan program and its area coverage goals.

The extremely remote area, encompassing 1,000 square miles in western North Dakota and nearby Montana, was one of the few remaining large areas of the United States without telephone service. Prior to Reservation's service installation, the residents communicated via citizens band and two-way radios, and these often did not work during the day due to noise and interference.

In early 1970, residents began exploring the possibilities of getting service from Northwestern Bell or a Montana cooperative or even building their own system. But costs proved prohibitive in all cases.

Then in May, several Squaw Gap residents met with the Reservation board. A week later, 85 people attended a public meeting held in Squaw Gap, and 70 of them signed $5 membership agreements. After an engineering study, Reservation offered to provide service for $12 a month, and everyone attending a July meeting voted to move ahead. Reservation eventually signed up 100 customers.

With one subscriber per 2.15 miles of line and a $4,600 investment per station, the project was not the usual installation. But the REA committed its funds and assistance to allow Reservation to serve the area. The 2 percent REA loan for $485,000 came through in November, but construction actually had begun a few months earlier.

The use of buried cable and the availability of new electronic components—as well as the dedicated efforts of the telephone company staff, the engineers, and the REA staff—made the project feasible. The automatic dial equipment was housed in an unattended building in Squaw Gap.

At 1:30 p.m. on December 15, 1971, many of the 100 Squaw Gap customers and various guests, as well as all four students from the Squaw Gap school, gathered in the community hall for the cutover. As part of the ceremony, Reservation President Donnell Haugen made the first phone call from Squaw Gap to Secretary of Agriculture Earl Butz in Washington, D.C. During the historic call, Butz observed, "This is one of the best examples of people helping themselves for better living with the aid of the government, and that's the way we ought to run the government and run this country."

The Squaw Gap success story made page one of *The Wall Street Journal* and also received national television coverage, letting everyone know just what the REA telephone program could accomplish.

The one-room Squaw Gap school, attended by four students, accurately portrays the remoteness of the service area.

NEWS

During the crisis in early 1972, many REA borrowers personally went to Washington, D.C., to discuss the situation with members of Congress. Visiting with Representative James Abdnor (R-SD) (seated) are (from left) Ransom Knutson, president of Consolidated Telephone Cooperative, Dickinson, North Dakota; Larry Wilson, board member of Grand Electric Cooperative; and Leroy Schecher, manager of West River Cooperative Telephone Company, Bison, South Dakota.

The REA's Darkest Day

Today Is

29

Friday December

Friday, December 29, 1972, the last business day of the calendar year. On that day, the Nixon administration terminated the existing REA program, and did so by simply issuing a routine U.S. Department of Agriculture press release.

The announcement stated that as of January 1, 1973, loans no longer would be made under the Rural Electrification Act, but rather under the Rural Development Act of 1972. Instead of direct loans, the REA would make insured and guaranteed loans, both at higher interest rates. Reportedly, neither REA Administrator David Hamil nor Secretary of Agriculture Earl Butz had advance notice of the decision.

Adding injury to insult, President Richard Nixon, in a televised news conference a short time later, stated that 80 percent of the 2 percent loans the REA made go "for country clubs and dilettantes, for example, and others who can afford living in the country."

Even before all the technical and procedural details of this new REA program had been addressed, the REA announced its first loans under the Rural Development Act in late February. By mid-May, the REA had made 24 loans, totaling approximately $40 million, and had another $50 million of loans in progress.

Grass-roots opposition mobilized in both the telephone and electric industries, as REA borrowers flew to Washington, D.C., and personally met with nearly every member of Congress. The lobbying efforts brought swift results—by mid-January, Representative Frank Denholm (D-SD) and Senators Hubert Humphrey (D-MN) and George Aiken (R-VT) had introduced measures calling for the restoration of the REA loan programs.

As word spread of the administration's intention to veto any such measure, Representative W.R. "Bob" Poage (D-TX) and Representative Ancher Nelsen (R-MN), former REA administrator, introduced a compromise measure which would create a revolving fund for REA lending. Congress refined this compromise into a form that eventually passed both the House and Senate.

While seeking to reinstate the program through legislative means remained the primary focus at the time, a side issue brewed as well: many viewed the administration's action as an impoundment of federal funds. A U.S district court judge ruled against the administration in a lawsuit filed on the matter.

Ending months of controversy about the future of the REA loan program, President Richard Nixon signs the bill restoring the program. Joining him are (from left) Assistant Secretary of Agriculture William Erwin, REA Administrator David Hamil, and Secretary of Agriculture Earl Butz.

The New Face of the REA

I n the end, the REA program re-emerged, but significantly changed, when President Richard Nixon signed an amendment to the Rural Electrification Act on May 11, 1973—coincidentally the 38th anniversary of President Franklin Roosevelt's executive order creating the REA program. The amendment established both insured and guaranteed loan programs, and also reduced the impact of the REA program on the federal budget, a goal many in the administration had long hoped to achieve.

The new Title III of the Rural Electrification Act created the Rural Electrification and Telephone Revolving Fund in the U.S. Treasury as a source for all REA loans. This eliminated the need for annual appropriations from Congress, as the fund was to be replenished by payments on outstanding and future REA loans and from the sale of borrower notes to the U.S. Treasury or the private money market. Congress did, however, retain the ability to set loan levels for each year.

A major change came in the area of REA interest rates. The rate for REA loans increased to 5 percent, with 2 percent monies reserved for special hardship cases, while the rate for RTB loans shifted to a single rate tied to the government's cost of money. The REA would use various criteria, including subscriber density and gross revenue, to determine the type of loan for which a borrower qualified. The amendment additionally authorized the REA to guarantee loans made by other lenders, but at the time, the provision really applied only to the electric program.

The 1973 amendment also lifted the requirement that RTB loans be made only to those who previously had borrowed from the REA, and the Boone County Telephone Company, Little Rock, Arkansas, became the first RTB-only borrower in September 1974. Also, the agency determined those who had received loans under the interim Rural Development Act-based program would have the option of converting those loans to REA loans.

In September 1975, the administration suggested several technical amendments to the act, specifically adjusting the eligibility criteria for the various types of loans and moving the $89 million in impounded funds to the revolving fund. Congress passed the measure in fall 1976.

On August 20, 1974, Federal Financing Bank (FFB) President Jack Bennett (center) signs the agreement whereby the FFB will make loans guaranteed by the REA. With him are Assistant Secretary of Agriculture William Erwin (left) and REA Administrator David Hamil (right).

A Guaranteed Option

With a large loan backlog still looming, the REA began to look at a guaranteed loan segment for the telephone loan program. Following the passage of the Federal Financing Bank (FFB) Act in December 1973, this entity became the REA's obvious choice. Congress had formed the FFB to serve as the single outlet to the private financial market for government agencies.

On August 20, 1974, the REA and the FFB signed an agreement whereby the FFB would provide funds for any loan for which the REA would provide a guarantee.

The REA made its first FFB telephone loan commitments to three companies on December 11, 1974: Doniphan Telephone Company, Doniphan, Missouri; St. Joseph Telephone and Telegraph Company, Port St. Joe, Florida; and West Virginia Telephone Company, St. Marys, West Virginia. It then made its first FFB loan advance to Doniphan Telephone Company on April 10, 1975, for $539,207.

With the addition of the FFB option and the other changes to the program under way, the REA saw its first significant reduction in the loan backlog during 1976.

REA Administrator David Hamil (seated) signs the first guaranteed loan for a telephone system on December 11, 1974. Joining him are (from left) Assistant Administrator - Telephone Ray Ballard; United States Independent Telephone Association (USITA) Executive Vice President William Mott; National REA Telephone Association General Counsel A. Harold Peterson; USITA President Hugh Wilbourn of Allied Telephone Company, Little Rock, Arkansas; REA Southwest Area Director for Telephone Bill Nichols; and National Telephone Cooperative Association Executive Vice President David Fullarton.

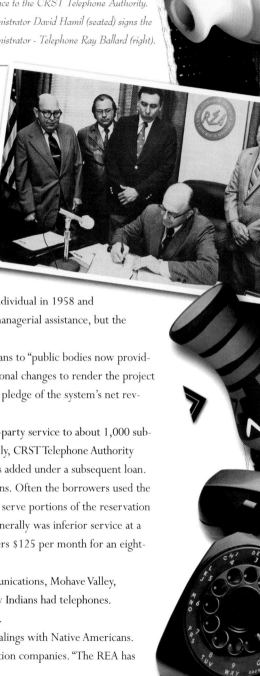

After many years of assisting the telephone company, the REA finally is able to provide financial assistance to the CRST Telephone Authority. Manager Dennis Calhoun (second from left) and President Darrel Hump (third from left) watch as REA Administrator David Hamil (seated) signs the loan. Also present are REA North Central Area Director William Riley (far left) and Assistant Administrator - Telephone Ray Ballard (right).

The REA on the Reservation

When the REA signed a loan for the Cheyenne River Sioux Tribe (CRST) Telephone Authority, Eagle Butte, South Dakota, on March 19, 1976, the agency registered a double first—not only was it the first loan to an Indian tribe, but it also was the first loan to a public body. But even more importantly than those publicized "firsts," the loan represented the agency's first of many successful partnerships with Native Americans.

Located in north central South Dakota, the CRST operated a telephone franchise it had purchased from an individual in 1958 and had been struggling to keep going. REA field personnel often had visited the reservation to offer technical and managerial assistance, but the agency was not authorized to make a loan to the system.

That situation changed, however, when a May 1971 amendment to the Rural Electrification Act permitted loans to "public bodies now providing telephone service in rural areas." The REA worked with the CRST to make several management and operational changes to render the project feasible. Also, because the tribe, like a municipal authority, could not mortgage its property, the REA arranged a pledge of the system's net revenues instead.

The CRST used the $3.2 million loan to upgrade service to the reservation, providing all underground, single-party service to about 1,000 subscribers, as well as to build a headquarters building in Eagle Butte and a new dial central office in LaPlant. Ironically, CRST Telephone Authority President Darrel Hump did not receive a telephone in the first upgrade; he would be one of the 1,000 subscribers added under a subsequent loan.

In the years since this landmark loan, the REA/RUS has made many more loans to Native American operations. Often the borrowers used the funds to upgrade systems they purchased from the Bell companies. These larger companies frequently refused to serve portions of the reservation or would charge exorbitant amounts to extend lines. And when they did provide service to the reservation, it generally was inferior service at a hefty price. Tohono O'Odham Utility Authority, Sells, Arizona, for instance, said U S West charged tribal members $125 per month for an eight-party line. Now, they pay only $10 per month for a single-party line.

"There wouldn't have been a company without REA," said Ken Doughty, president of Fort Mojave Telecommunications, Mohave Valley, Arizona. "We got a $2 million loan and started from scratch." At the time of the loan in the early 1990s, very few Indians had telephones. Now, the tribe owns the majority stake in the telephone company and offers subscribers many advanced features.

Across the board, the reservation telephone systems commend the REA/RUS on its fairness and unbiased dealings with Native Americans. Many credit the REA's treatment of the CRST as the model that laid the groundwork for future loans to reservation companies. "The REA has always been sensitive to Indian issues," said Doughty.

Gary Hadfield of Bristol Bay Telephone Cooperative, King Salmon, Alaska (left), repairs a subscriber's telephone set. In remote areas of the state where roads often are non-existent, having a telephone can be a matter of life and death.

Conquering the Last Frontier

L ooking to the far north, the REA found Alaskan natives to be severely lacking when it came to telephone service. "As late as the '70s, many villages had only one public phone," remembered Marie Green, a board member of the OTZ Telephone Cooperative, Kotzebue, Alaska. "And there were major celebrations when that one phone was installed."

OTZ took the spotlight as one of the REA's big success stories in the mid-1970s when it used a $3.4 million loan to bring single-party service to 1,000 subscribers in 10 villages scattered in a 36,000-square-mile area largely above the Arctic Circle. Prior to the OTZ project, only three telephones existed in all the villages combined, and these were bush telephones connecting through Nome.

In June 1975, two REA engineers traveled to the region to conduct a preloan study. After evaluating several possible system designs, the REA and OTZ decided to use satellite and microwave to link the villages to the outside world through Kotzebue. Seven of the 10 villages, plus Kotzebue, had their own central offices. Many of the buildings were constructed in Seattle and transported to the region via barge.

Due to the extreme climate and terrain, the REA found itself dealing with an entirely new set of construction challenges. With temperatures averaging in the teens and often reaching far below zero, construction crews could work outside only three months of the year. And due to the permafrost, cable could not be buried. Instead it was placed along the ground in long rectangular wooden boxes called utilidors. Supplies could be flown in only when the weather was good or shipped on one of the two barges that came to the area each year.

After completing the project in 1977, OTZ continued to work to improve service in rural western Alaska. Over the years, loans from the REA/RUS allowed OTZ to hook up additional villages and further improve service. The cooperative now serves 3,200 people, and its most recent RUS loan, finalized in July 1998, will allow it to install signaling system 7 (SS7) technology and offer Caller ID.

Realizing the unique tasks of providing telephone service in the 50th state would continue to challenge rural telephone systems, the REA in August 1977 awarded a $70,000 contract to Edutel Communications and Development of Washington, D.C., to study potential communications technology solutions for rural Alaska. But still, the lack of rural Alaskan communications systems continued into the 1980s and even the 1990s.

Prior to the OTZ project, citizens of Noorvik, Alaska, including these young Eskimo girls, did not have telephone service.

On December 27, 1974, the REA granted Lafourche Telephone Company (Latelco), Larose, Louisiana, a $1.68 million loan, making possible the installation of the world's first fully stored program electronic switching system designed for smaller communities. The ITT TCS-5 electronic switch arrived in Galliano, Louisiana, in April 1976, and on June 14—just 57 days later—Latelco cut over to the new system. The installation, shown here, progressed so quickly because the switch had been wired, assembled, and tested under field conditions at the ITT factory in Illinois. The new Galliano office consolidated traffic from three separate offices into one switch. The electronic switch took up only one-third of the space occupied by the step-by-step equipment it replaced, and it allowed the company to offer customers call forwarding, call waiting, three-way calling, and speed dialing.

Picking Up the Pace

The rate of technological developments increased in the 1970s, and not just the REA, but the entire telecommunications industry, began to explore the possibilities of digital switching, fiber optics, and broadband applications. Continuing its innovative engineering approach, the REA established a Future Network Development Branch, and its work in these areas resulted in several major milestones for the agency.

Borrowers increasingly replaced their electromechanical central office switches with common control electronic switches. Digital switching came on the scene in 1976, and by 1979, the REA was finding the majority of its borrowers' central office replacement studies favored installing a digital switch.

Throughout the decade, the oil crisis resulted in a plastic cable shortage, which drove up materials costs. One REA technical staff member recalled a time when telephone cable costs increased 30 percent in a three-month period. The REA's deficiency funding program—allowing borrowers to obtain additional funds if the initial loan amount was inadequate—proved key to borrowers' surmounting such obstacles. But ironically, the situation had a positive angle. Due to the lack of cable, the REA re-evaluated its design, installation, and operating criteria to increase the use of electronics and reduce the amount of cable. This new system approach brought an overall cost savings to borrowers.

With the growth of new technologies, the REA realized borrowers would have to deal with faster obsolescence. Thus, the agency began encouraging borrowers to prepare long-range plans. In 1977, the agency published a booklet titled *Serving Telephone Subscribers in the Year 2000*. The REA also implemented its Service Area Value Engineering (SAVE) concept. These guidelines assisted borrowers in preparing the optimum system design—one that was not too costly to install but still had enough capacity for the future; one that used the right combination of copper and electronics. SAVE became an established requirement on July 1, 1977.

The REA stepped up its efforts to move larger borrowers toward greater self-sufficiency. In late 1977, the agency approved construction certification procedures to allow larger borrowers to approve their own plans, specifications, bids, contracts, and inspections, thus reducing the paper and workload at the REA.

To adequately assist borrowers with the many industry changes, the REA created its Telecommunications Management Division and offered borrowers additional marketing and sales assistance. The agency also increased its work with other standards-setting bodies in the industry.

But by the late 1970s, a trend began to emerge that would remain evident for nearly two decades—the industry's concerns about the administration's management of—and future plans for—the REA program, and Congress' actions to address those concerns. Among the primary issues: the apparent need for the agency to hire additional personnel in the telephone program, particularly with the growth of new telecommunications technologies.

Following the cutover of an eight-mile link of coaxial cable at Chequamegon Telephone Cooperative, Cable, Wisconsin, in 1076, those making this broadband voice project possible share in the first telephone calls over the facilities: (from left) REA Assistant Administrator - Telephone Ray Ballard in Washington, D.C.; 3M Group Vice President E. Wayne Bollmeier and 3M TelComm Products Department Manager Gary Pint in St. Paul, Minnesota; and Chequamegon General Manager Alan Olson.

Broadband Before its Time

True to its track record of innovations and staying on the leading edge of new technologies, the REA emerged on the forefront as applications of broadband technology began to unfold. But this time, unfortunately, the agency found its approach to be a bit ahead of its time.

Even in the late 1960s, REA engineers already were looking at how to provide total communications—telephone, video, computer access, alarm services, and more—using the same coaxial cable. The agency presented a paper, *A Challenge and a Blueprint for a Total Telecommunications System,* at a 1969 symposium.

Then in October 1977, the agency contracted Bell Northern Research in Palo Alto, California, to study the technical and economic feasibility of providing narrowband and broadband communications services in rural areas. The study looked at both fiber optic and coaxial cable.

The REA began encouraging the development of broadband equipment that would allow the simultaneous transmission of telephone, television, and other communications services via the same coaxial cable. The equipment that the REA thought could accomplish its goal: the 3M Coaxial Subscriber System (CS²); it had worked well in the laboratory and needed only to prove itself in the field.

In August 1978, the agency approved a $1.4 million loan to partially finance a demonstration project for the 3M system at Footville Telephone Company, Footville, Wisconsin. Because the REA could not lend for commercial television service, it could finance only a portion of the project costs. But as the project began to move forward, it faced several obstacles related to the cable television portion, including some regulatory delays. Garden Valley Telephone Company, Erskine, Minnesota, also field trialed the system and encountered problems with the functioning of the television part of the system as well.

Meanwhile, Chequamegon Telephone Cooperative, Cable, Wisconsin, had a similar project under way. The REA in 1977 had approved the first of two loans for the cooperative's system upgrade, which included a two-phase coaxial cable project. Phase one offered voice service, while phase two—the long-term goal—was to offer cable television service over the same facilities. Chequamegon also installed the 3M system.

But much to the chagrin of the REA and its borrowers, the system never worked as planned. While the voice application performed fine, the integrated services aspect never functioned in the field. After several years and lots of dollars, 3M eventually discontinued manufacture of the system, citing technological problems. Thus ended the REA's early foray into the broadband world.

Prior to making the first call via an REA-funded fiber optic system, key project participants check the equipment: (from left) Commonwealth President William Umphred, REA Northeast Area Chief Engineer William Bird, and Kenneth Ray of ITT.

The REA's Fiber First

An agreement executed on June 8, 1978, between the REA and Commonwealth Telephone Company, Dallas, Pennsylvania, marked a milestone not only for the agency and rural telephony, but also for the entire U.S. telecommunications industry. The agreement was for the design and construction of the nation's first fiber optic system for commercial use.

Working with REA engineers, as well as system manufacturer ITT, Commonwealth installed a 13.5-mile fiber system during an eight-week period. The system supplemented existing overloaded circuits on the company's toll route between Mansfield and Wellsboro in north central Pennsylvania. The fiber cable could carry 672 simultaneous conversations—more than six and a half times the capacity of the standard copper cables of the day.

The project proved unique in that Commonwealth installed about half of the fiber as buried plant and the other half as aerial plant. The trial included four methods of construction—direct burial, underground duct, aerial lashed to existing cable, and aerial lashed to new messenger—allowing the REA to obtain data about all types of installations. While all techniques proved equally effective, those who worked on the project recall that the buried portion appealed to the tastes of the local groundhogs, which resulted in an outage not long after the installation.

As Commonwealth crews used their standard ditching plows to bury the fiber, they found themselves trying to be extra careful. But after about the first kilometer, they realized it could be handled just like standard cable and actually was easier to maneuver. They effortlessly hand-pulled the duct portion of the installation, and as for the lashing, the lightweight fiber cable allowed the process to move quite quickly.

Recognizing the benefits of the project to all of rural telephony, the REA bore the additional cost of the fiber components above that of conventional construction. And indeed, the installation provided extensive information in the areas of fiber placement and splicing, as well as on the effects of the environment on such a system. Although the REA classified the installation as a field trial, the system provided fully functional permanent service to Commonwealth's subscribers.

During the year, the telephone company played host to many key telephone industry players and government officials from throughout the United States, as well as from a few foreign countries. Also, ITT produced a film based on the project titled *Fiber Optics: The Light That Talks*. Commonwealth had in operation one of the longest fiber optic systems in the world, as well as one of the first systems to use optical repeaters.

On June 7, 1979, Assistant Secretary of Agriculture Alex Mecure receives the first commercial fiber optic telephone call via Commonwealth's facilities from President William Umphred.

During a celebration of the REA telephone program's 25th anniversary in October 1974, current administrator David Hamil (center) is joined by former administrators Ancher Nelsen (left) and Norman Clapp (right). Hamil, who served two terms as administrator, has been called "Mr. REA" by many and is specifically recognized for operating the REA program in a non-partisan atmosphere throughout his tenure. When he left office in 1078, he had approved more than 70 percent of the total REA telephone financing.

Industry Involvement

By the mid-1970s, decisions at the Federal Communications Commission (FCC) had begun to have a greater impact on REA telephone borrowers, and the agency found itself expanding its role to include such matters.

Much FCC activity during this time came in the area of telephone equipment, as the commission determined subscribers could provide their own. The commission undertook an extensive equipment registration program, and the REA, in conjunction with the National Telephone Cooperative Association and the other telephone trade associations, developed a list of items for the "grandfather" provision of the program. They asked that the list be accepted in lieu of an individual list from each telephone company. The FCC also adopted standard plugs and jacks for connecting certain types of equipment, and the REA worked with manufacturers to ensure adequate quantities would be available for its borrowers.

To assist borrowers in managing these changes, the REA added a telecommunications marketing specialist to its staff in 1976 and developed a telecommunications marketing handbook. The agency focused on expanding vertical services and on EPIC—equipment partially installed by customer—encouraging borrowers to move to modular telephone sets and establish telephone stores to display equipment to their subscribers.

The following year, the REA added a separations specialist to its staff to assist borrowers in determining whether cost or average schedule settlements proved more beneficial to their operations.

As Congress considered changes to the Communications Act, borrowers had growing concerns about the impact of competition and toll rate deaveraging. Other regulatory initiatives on the state level affected borrowers as well. While some states' environmental concerns pushed them to require underground plant, others implemented budget service for low-income and elderly subscribers.

REA Administrator Robert Feragen (center), Assistant Administrator - Telephone John Arnesen (left) and North Central Area Director William Riley (right) participate in the cake cutting during a reception honoring the 30th anniversary of the REA telephone program in 1079.

Lending for Cable

In May 1979, the REA entered a new, albeit related, industry. As part of President Jimmy Carter's initiative on improving rural communications, Secretary of Agriculture Bob Bergland transferred authority for making loans and guarantees for cable television facilities from the Farmers Home Administration to the REA. On September 27, 1979, REA Administrator Robert Feragen signed the first two loans, made under the Consolidated Farm and Rural Development Act.

One was a $3 million loan to Great Plains Broadband Services, Kingfisher, Oklahoma, to finance construction of a system to serve more than 3,400 rural subscribers via 412 miles of coaxial cable. The other, a 90 percent guarantee of a $2.9 million loan to Tel-Com Inc. of Harold, Kentucky, would provide or improve reception for more than 6,400 homes in Kentucky and parts of West Virginia.

As they had in both the electric and telephone loan programs, the REA staff set about developing system designs and engineering guidelines specifically geared to rural cable television applications. However, the REA's adventure in cable television lending ended shortly after it began when the fiscal year 1982 budget did not include funding for the program. The agency did, however, continue to maintain the loans it had made.

REA Administrator Robert Feragen (center) signs a $3 million loan for Great Plains Broadband Services, Kingfisher, Oklahoma. Joining him are (from left) Secretary of Agriculture Bob Bergland, Great Plains President Marion Perdue, and Great Plains Manager Johnnie Ruhl. Also attending the ceremony was Paul Gearhart, president and manager of Tel-Com Inc., Harold, Kentucky, the other company receiving one of the initial REA cable television loans.

Facts and Figures:
Rural Telephone Bank Directors

Since its creation in 1971, the Rural Telephone Bank (RTB) board of directors has included six industry representatives. Three of the representatives are from telephone cooperatives, while the other three are from commercial telephone companies. The following are the individuals who have served as industry directors of the RTB.

Glenn W. Bergland, *Winnebago Cooperative Telephone Association,*
Thompson, Iowa*.. 9/71 - 8/74
Lake Mills Telephone Company, Lake Mills, Iowa .... 9/74 - 9/80

Jean S. Brandli, *Coosa Valley Telephone Company,*
Pell City, Alabama... 9/71 - 9/74

Leroy T. Carlson, *Telephone and Data Systems,*
Chicago, Illinois .... 9/71 - 9/72

B. Maynard Christenson, *Dakota Cooperative Telephone Company,*
Irene, South Dakota* .... 9/71 - 12/74

David J. McKay, *Golden Belt Telephone Association,*
LaCrosse, Kansas .... 9/71 - 9/72

Harold G. Payne, *Telephone Utilities of Pennsylvania,*
Export, Pennsylvania .... 9/71 - 9/80

Roy C. Boecher, *Pioneer Telephone Cooperative,*
Kingfisher, Oklahoma*.. 9/72 - 9/76

Arndon O. Haynes, *Mashell Telephone Company,*
Eatonville, Washington .... 9/72 - 2/92

John A. McAllister, *West Carolina Rural Telephone Cooperative,*
Abbeville, South Carolina* .... 9/74 - present

Harold C. Ebaugh, *Triangle Telephone Cooperative Association,*
Havre, Montana*.. 2/75 - 9/76

The first Rural Telephone Bank board of directors, appointed by President Richard Nixon in 1971, includes (from left) Clarence Palmby, former assistant secretary of agriculture; Harold Payne, Telephone Utilities of Pennsylvania, Export, Pennsylvania; B. Maynard Christenson, Dakota Cooperative Telephone Company, Irene, South Dakota; Jean Brandli, Coosa Valley Telephone Company, Pell City, Alabama; Leroy Carlson, Telephone and Data Systems, Chicago, Illinois; Dr. Thomas Cowden, Assistant Secretary of Agriculture; E.A. Jaenke, governor of the Farm Credit Administration; REA Administrator David Hamil; Dr. Eric Thor, administrator of the Farmer Cooperative Service; Willis Ward, chairman of the Michigan Public Service Commission; Charles Pillard, International Brotherhood of Electrical Workers; David McKay, Golden Belt Telephone Association, LaCrosse, Kansas; and Glenn Bergland, Winnebago Cooperative Telephone Association, Thompson, Iowa.

Representing the first five RTB borrowers at the January 21, 1972, loan signing are (from left) Joseph Chilen, manager of Pioneer Telephone Association, Ulysses, Kansas; Taft Maddox, manager of Eastex Telephone Cooperative, Henderson, Texas; Earl Williams, president of Pioneer; Ross Vernon, manager of Mutual Telephone Company, Sioux Center, Iowa; RTB Governor David Hamil; Richard Rozeboom, president of Mutual; William Fitzgerald, president and manager of Randolph Telephone Company, Liberty, North Carolina; Maurice TePaske, attorney for Mutual and mayor of Sioux Center; J.P. Gleaton, president of Plant Telephone and Power Company, Tifton, Georgia; and Fred Bailey, vice president and general manager of Plant.

Glenn Nelson, *Farmers Mutual Telephone Company,* Bellingham, Minnesota* .... 9/76 - 9/78

Maurice Woodward, *Hancock Rural Telephone Corporation,* Maxwell, Indiana* .... 9/76 - 8/77

Charles M. Means, *South Plains Telephone Cooperative,* Lubbock, Texas/
Hill Country Telephone Cooperative, Ingram, Texas* .... 11/77 - 9/88

Eldon M. Snowden, *McDonough Telephone Cooperative,* Colchester, Illinois* .... 9/78 - 9/84

Eleanor Haskin, *Waitsfield-Fayston Telephone Company,* Waitsfield, Vermont .... 9/80 - 9/88

W.S. "Babe" Howard, *Millington Telephone Company,* Millington, Tennessee .... 9/80 - 9/88

Joseph Vellone, *National Telephone Cooperative Association,* Washington, D.C.* .... 9/84 - 11/94

James H. Creekmore, *Delta Telephone Company,* Louise, Mississippi .... 9/88 - 9/90

Gary Kennedy, *Panhandle Telephone Cooperative,* Guymon, Oklahoma/
Telecommunications Consultant, Oklahoma* .... 9/88 - 11/98

Curtis A. Sampson, *Communications Systems/
Hector Communications Corporation,* Hector, Minnesota .... 9/88 - present

Jeff L. McGehee, *Southland Telephone Company,* Atmore, Alabama .... 9/90 - 2/92

Jack L. Bentley, *Western New Mexico Telephone Company,* Silver City, New Mexico/
Telecommunications Consultant, New Mexico .... 2/92 - 11/98

S. Michael Jensen, *Great Plains Communications,* Blair, Nebraska .... 2/92 - present

Larry Sevier, *Rural Telephone Service Company,* Lenora, Kansas* .... 11/94 - present

David Crothers, *North Dakota Association of Telephone Cooperatives,*
Mandan, North Dakota* .... 11/98 - present

John Dillard, *Monroe Telephone Company,* Monroe, Oregon .... 11/98 - present

* indicates cooperative representative

Reviewing the measure giving the RTB the authority to obtain funds by selling its debentures to the U.S. Treasury are (from left) REA Deputy Administrator Everett Weitzell, Administrator David Hamil, and Assistant Administrator - Telephone Frank Renshaw.

UNITED STATES DISTRICT COURT
FOR THE DISTRICT OF COLUMBIA

UNITED STATES OF AMERICA,

                          Plaintiff,   )
          v.   )
)
WESTERN ELECTRIC COMPANY,   )   Civil Action No.
INCORPORATED, AND AMERICAN   )   82-0192
TELEPHONE AND TELEGRAPH   )
COMPANY,   )
)
                          Defendants.   )

MODIFICATION OF FINAL JUDGMENT

Plaintiff, United States of America, having filed its
complaint herein on January 14, 1949; the defendants having
appeared and filed their answer to such complaint denying the
substantive allegations thereof; the parties, by the...
attorneys, having severally consented to...
was entered by the Court on Jan...
having subsequently a...
Judgment is...

FILED
AUG 24 1982
JAMES F. DAVEY, Clerk.

A s the telephone industry entered the 1980s, it witnessed the advent of deregulation and competition and the implementation of divestiture and access charges. While these industry events generally were not supposed to impact independents, in reality, they greatly changed borrowers' operations and added a new dimension to the REA's functions. Meanwhile, controversy surrounding the future existence of the REA loan program grew stronger than ever. Still, the REA staff continued their hard work and accomplished much for rural telephony and rural America. For both the agency and its borrowers, the 1980s truly brought many ...

Questions and Concerns

The REA, as well as its borrowers, explored the possibilities of satellite technology for bringing services to rural areas.

By the 1980s, REA borrowers had single-party service in place for the majority of subscribers and thus used REA financing to continue to expand and modernize their facilities. Here, Clarks Telephone Company, Clarks, Nebraska, installs new toll facilities in its service area in 1982. The company used REA/RUS financing in 1959 for its step-by-step switch, in the mid-1970s to move from eight-party to one-party service, and now in the late 1990s to upgrade its central office switch to provide intraLATA equal access.

The Bigger Picture

B y the 1980s, the REA had become even more involved in general telecommunications industry affairs, filing comments with the Federal Communications Commission (FCC), offering testimony on Capitol Hill, and determining what the numerous legislative and regulatory decisions meant for REA borrowers. REA regional seminar topics focused not only on developing technologies, but also on issues such as new FCC regulations and toll settlement changes. In 1986, the REA and FCC signed an agreement to assist one another in monitoring the effects of federal decisions on rural telephone companies.

By far, the biggest event of the decade—indeed in the history of the telecommunications industry to that point—came on January 1, 1984, when the court-ordered divestiture of the Bell companies from AT&T went into effect. While divestiture was supposed to be "transparent" to the independent telephone industry, it actually changed nearly every aspect of their operations.

Among the many regulatory matters tied to divestiture was the FCC's access charge decision. When the commission issued its original order, the REA conducted a study on its impact on borrowers' rates and concluded that rates would increase by at least 100 percent and as much as 300 percent or more by the end of the established transition period. The REA not only testified before Congress, but also produced a videotape on the issue for the REA field staff to show to telephone company managers and boards of directors.

Borrowers also faced the changes brought by the FCC orders deregulating telephone equipment and inside wiring. To assist with the confusion resulting from these FCC decisions, the REA issued a publication explaining the issues.

Other matters needing the REA's attention included the FCC's new Uniform System of Accounts, rulings on equal access for toll carriers, and the telephone company-cable television cross-ownership issue, as well as Congress' continued attempts to rewrite the Communications Act of 1934.

As borrowers started to diversify their operations, the REA found itself reviewing proposals and contracts for creating subsidiaries and affiliates in areas such as resale, regional toll networks, cable television, satellite dishes, cellular service, and electronic equipment. The agency also helped borrowers conduct financial feasibility studies to determine their eligibility for cellular markets.

Federal Communications Commission   213557

FCC 87-387

Before the
Federal Communications Commission
Washington, D.C. 20554

CC Docket No. 86-495

In the Matter of

Basic Exchange
Telecommunications   RM 5442
Radio Service

REPORT AND ORDER

Adopted: December 10, 1987; Released: Jan...

By the Commission...

Art Biggs, president of S&A Telephone Company, Allen, Kansas, along with a subscriber, inspects the BETRS equipment installation at the subscriber's home.

Two Innovative Ideas

As the REA continued to evaluate developing technologies and their application to rural telecommunications, it undertook two major projects during the 1980s, both in conjunction with the National Telephone Cooperative Association (NTCA) and other industry trade associations.

The first project developed from the agency's support of NASA's 1983 petition to the Federal Communications Commission (FCC) seeking a spectrum allocation for mobile satellite service. The REA began to look at satellite technology and determined three services had potential in rural areas: rural satellite toll trunking, rural satellite subscriber loops, and rural satellite mobile radio service. After the REA and interested rural trade associations heard 12 presentations in summer 1983, they decided to pursue toll trunking and established subcommittees to study the various aspects of a national rural telephone toll network using satellite technology.

Basically, the proposal called for independent telephone systems to launch a satellite and use it to carry their own long distance traffic. After completing a sample rural toll traffic survey and distributing it to interested manufacturers, the REA and the associations listened to specific proposals from Hughes Communications, GTE Spacenet, and Ford Aerospace.

But after evaluating the costs, the coalition determined the system was not feasible for the REA to finance and borrowers to own and operate. Instead, the REA recommended rural companies consider developing regional terrestrial toll networks, which many small companies later did.

In May 1986, the REA again joined the NTCA and other national telephone trade associations, this time in petitioning the FCC to establish basic exchange telecommunications radio service (BETRS). The service would use designated frequencies on a shared basis with other services to provide wireless telephone service in remote areas.

The REA viewed BETRS as an alternative to wireline service for both new service and multiparty service upgrades in remote areas, as well as a solution for emergency or seasonal service. The FCC approved the petition in late 1987, and by the end of summer 1988, REA borrower S&A Telephone Company, Allen, Kansas, satisfactorily concluded a field trial of the first BETRS equipment to be added to the REA's list of approved materials.

While BETRS never achieved the level of use the REA originally had anticipated, it did bring service to many remote subscribers.

The Battle of the Budget

When President Ronald Reagan took office in January 1981, his administration promptly presented the first of eight annual budget submissions that each sought to severely cut—or even eliminate—the REA electric and telephone loan programs. The proposals ranged from immediate termination of the program to a variety of phase-out approaches, such as replacing the program with a partial private loan guarantee program that not only would carry a higher interest rate, but for which the majority of borrowers would not be eligible.

But despite the repeated attacks, congressional support for the REA program remained strong. Year after year, the administration failed to find members of Congress willing to introduce its anti-REA packages. And due to the hard lobbying work of borrowers and their trade associations, Congress retained the lending levels for the REA programs throughout the Reagan years. Unfortunately, however, constant loan levels actually translated into less real dollars as each year passed.

When President George Bush entered the Oval Office in 1989, the telephone industry hoped for a "kinder, gentler" approach to the REA budget, particularly since a plank in the 1988 Republican party platform pledged continued support for the REA. But Bush's proposals did not appear much different than his predecessor's.

Congress continued to maintain loan levels, but in 1989, the nation faced a Gramm-Rudman-Hollings sequestration order, effective through February 1990. The measure cut all domestic spending by 1.7 percent, including REA funding. While disappointed with the drop in available funding, REA borrowers accepted the action as something all federal programs had to endure.

Another budget dilemma impacted the REA program in 1990 when Congress ordered all federal programs cut by 25 percent. For the REA, the 25 percent reduction was transferred to a 90 percent private guaranteed loan program. When the fiscal year 1991 budget again called for REA funding to be at the previous year's levels, the amounts were 25 percent less due to the cut. The appropriations measure, fortunately, restored funding to the previous levels, again due to the continued efforts of REA borrowers and their trade associations on Capitol Hill.

Following his breakfast speech at the 1981 National Telephone Cooperative Association (NTCA) Fall Conference in Boston, REA Administrator Harold Hunter (left) pauses to answer questions from President Earl Waterland (center) and General Manager Donald Paulsen (right), both of Golden West Telephone Cooperative, Wall, South Dakota. During his speech, Hunter discussed "new directions" at the REA, directions which did not prove popular with many NTCA members.

An Uncertain Future

In testimony before the House Conservation and Credit Subcommittee in April 1986, REA Administrator Harold Hunter put forth the administration's view: "The REA's mission has been accomplished." Lacking the political support to significantly modify the REA via the budget process, the administration tried another route to phase out the program—policies and eligibility standards that made it difficult for borrowers to qualify for loans.

One such issue was the REA's general funds policy, which basically limited the amount of money a borrower could have in the bank and still be eligible to borrow from the REA. The agency said if a borrower's cash or other investments exceeded 8 percent of its total plant, the borrower must foot a portion of the bill for the project. The basic policy was not new; the REA instituted it in 1962 as a means of preventing abuses of loan funds. But by the 1980s, the policy had become a strict rule and when applied to loans and loan advances, prevented a large majority of borrowers from securing funds.

The administration also began suggesting that the REA's technical standards-setting functions no longer were necessary due to the increased role other industry organizations played in this area. It called for the elimination of the REA's Telecommunications Engineering and Standards Division. In 1985, Congress held a hearing on the proposal, which led to a General Accounting Office (GAO) study of the REA's standards-setting function, as well as its general funds policy. The GAO did not support the REA's actions and, in a follow-up opinion, found the agency applying its bulletins as actual rules in violation of the Administrative Procedures Act.

As the REA continued to implement such rules and policies, and borrowers and their trade associations continued to carry their concerns to the Hill, Congress held hearings on what it called the REA's "hip pocket" rules. As a result, the Omnibus Budget Reconciliation Act in December 1987 directed the REA to codify all of its rules and policies. During the next several years, the REA published its various bulletins and policies as proposed rules in the *Federal Register*, accepted industry comments on them, and then published final rules. But while many policies improved, several troublesome provisions remained. Congress also included language in the funding measures for fiscal years 1986 through 1989 prohibiting the REA's use of its general funds policy to deny access to loans.

Meanwhile, the industry continued to raise concerns about the limited REA staff, particularly in the technical and standards areas. Congress again let its voice be heard, prohibiting the REA from using appropriated funds to change the Telecommunications Engineering and Standards Division and directing the agency to start hiring to fill the many staff vacancies.

National Telephone Cooperative Association President Cheryl Borth, Consolidated Telephone Cooperative, Dickinson, North Dakota, testifies before the Senate Appropriations Subcommittee on Agriculture, Rural Development, and Related Agencies in April 1989 concerning the annual REA loan program appropriations.

Sheldon Petersen (standing), who became RTFC's CEO in 1995, speaks at RTFC's 1992 Annual Meeting as the staff administrative coordinator. He is joined at the table by two of RTFC's officers at that time: President Jim Wilson, president of XIT Rural Telephone Cooperative, Dalhart, Texas; and Secretary-Treasurer Hobart Davis, general manager of Skyline Telephone Membership Corporation, West Jefferson, North Carolina.

Sheldon Petersen (standing), who became RTFC's CEO in 1995, speaks at RTFC's 1992 Annual Meeting as the staff administrative coordinator. He is joined at the table by two of RTFC's officers at that time: President Jim Wilson, president of XIT Rural Telephone Cooperative, Dalhart, Texas; and Secretary-Treasurer Hobart Davis, general manager of Skyline Telephone Membership Corporation, West Jefferson, North Carolina.

Increasing Borrower Options

A nother matter of concern for borrowers was the REA's policy of retaining the first mortgage not only on the assets purchased with REA funds, but also on all assets previously owned and acquired thereafter. Such a policy made it nearly impossible for a borrower to obtain funding from any other source.

As borrowers began to diversify into areas other than basic telephone service and needed funding for purposes other than those authorized by the Rural Electrification Act, this issue grew in importance. Another catalyst for REA action on the matter was a December 1985 amendment to the Farm Bill which allowed those eligible for REA funding to borrow from the Farm Credit System Banks for Cooperatives (now CoBank), as well as the National Cooperative Bank.

The REA began work on a new lien accommodation and subordination policy and in September 1986, issued its first lien accommodation regulation, establishing a means for borrowers to attract capital from alternative sources.

An additional opportunity for such financing came into existence shortly thereafter when leaders of the rural telephone industry and the National Rural Utilities Cooperative Finance Corporation (CFC), the financing organization created by the electric cooperatives in 1969, joined together to form a new cooperative corporation to provide financing to the rural telecommunications industry. The Rural Telephone Finance Cooperative (RTFC) was incorporated in September 1987, with membership open to systems eligible to receive REA loans and their affiliates.

These alternative financing sources have become an integral part of the rural telecommunications industry, providing funding that allows for the continued improvement of telecommunications services and infrastructure in rural America. The RTFC, for instance, has nearly $3 billion of loans outstanding to rural telecommunications service providers and anticipates that strong demand for such financing will continue.

Several RTFC staff members visit XIT Rural Telephone Cooperative, Dalhart, Texas, to receive an orientation about telephone cooperative operations and new lines of business.

The RTB Moves Forward

A s the decade began, the Rural Telephone Bank (RTB) faced the end of its initial 10-year capitalization program. The RTB directors, as well as Congress, realized that if the government support halted, RTB rates would escalate, rendering many small companies unable to meet the RTB loan criteria and thus forcing them back to the lower-rate insured program. Legislation to extend the capitalization passed in 1981, calling for the government to continue to invest $30 million annually for an additional 10 years.

The bank also had several policy changes during the 1980s, including the establishment of a formula for RTB interest rates. But the primary RTB issue proved to be the pending privatization of the bank. Many in the administration began pushing for early privatization, but Congress precluded such action by prohibiting the RTB from creating reserves for retiring Class A stock.

A controversial RTB action came in November 1988, when the RTB board of directors authorized $400,000 be spent on a privatization study. Despite protests from the industry representatives on the RTB board, as well as from several members of Congress, the RTB hired Kidder, Peabody & Co. to conduct a 100-day study and develop a plan to privatize the bank in 1995. The six elected directors voted against the proposal and filed a protest with the General Accounting Office.

The Rural Telephone Bank (RTB) board of directors and several members of the REA telephone staff celebrate the RTB's 10-year anniversary in 1981. Deputy Under Secretary of Agriculture Ruth Reister cuts the cake, which was baked by Assistant Administrator - Telephone John Arnesen (second from left) and his wife, Alice (not pictured). The telephone industry RTB board members include (front row starting fourth from left) W.S. "Babe" Howard, Millington Telephone Company, Millington, Tennessee; Arndon Haynes, Mashell Telephone Company, Eatonville, Washington; Eleanor Haskin, Waitsfield-Fayston Telephone Company, Waitsfield, Vermont; John McAllister, West Carolina Rural Telephone Cooperative, Abbeville, South Carolina (standing behind); and Eldon Snowden, McDonough Telephone Cooperative, Colchester, Illinois. Charles Means, South Plains Telephone Cooperative, Lubbock, Texas, is not pictured.

More Modifications

The REA bid goodbye to one of its original features in August 1981 when amendments to the Rural Electrification Act eliminated 2 percent REA loans. The amendments did, however, permit exceptions to the 5 percent rate for hardship cases at the discretion of the administrator.

Other fundamental changes in the program's operations came during the 1980s, including a shift to monthly, instead of quarterly, loan payments for all new loans, beginning in late 1982, as well as a cut in the principal deferment from three years to two years. The same year, the REA said it no longer would make loans for headquarters facilities, terminal equipment, or company vehicles, pointing to recent Federal Communications Commission decisions as the basis for the change.

Starting in 1982, the REA underwent several years of constant staff adjustments. The agency combined its field engineer position and operations field representative position into a single general field representative position. The REA also rotated the area office leadership so employees could gain experience in other areas. Even the assistant administrators for the telephone and electric programs switched places in 1987.

Also, in an effort to reduce borrower dependency upon the REA programs, congressional spending measures throughout the late 1980s allowed the prepayment of various Federal Financing Bank (FFB) and Rural Telephone Bank (RTB) loans without penalty, and direct and insured loans at a discount. Congress also established a privatization demonstration program in Alaska, allowing borrowers in that state to prepay all FFB loans without penalty and also prepay all REA and RTB loans within the year if they agreed not to come back to the REA for future financing.

To allow it to place a higher degree of reliance on the financial information borrowers submit, the REA in 1988 launched its certified public accountant (CPA) peer review audit program. In doing so, the REA became the first federal agency to require CPAs to belong to a peer review program.

In January 1986, President Ronald Reagan signed the Compact of Free Association Act, which made foreign trust territories eligible for REA financing. With loans to the Federated States of Micronesia and the Marshall Islands, both actual nations, the REA truly became an international lender.

GRAND OPENING

Federated Telephone Cooperative, Chokio, Minnesota, was among the early Rural Economic Development Loan recipients and used one of its loans to construct a warehouse and purchase equipment for this seed cleaning facility at a cooperative grain elevator in Milan, Minnesota.

Reviving Rural Development

I n the same spirit as the Rural Areas Development (RAD) efforts launched in the 1960s, the federal government in the late 1980s developed new programs to assist REA/RUS borrowers in promoting the economic growth and development of their communities.

A December 1987 amendment to the Rural Electrification Act established the Rural Economic Development Loan and Grant programs to provide zero interest rate loans and grants to REA/RUS borrowers for promoting rural economic development and job creation projects. The funds go to the borrower, who in turn lends them to a specific development project. The project possibilities are virtually endless: constructing buildings or purchasing equipment for new businesses, renovating business districts, building recreational facilities, and even purchasing firetrucks.

For the loan portion of the program, zero interest rate loans are made to the borrower, who then makes a zero interest rate loan to the project. The grant portion of the program is more restrictive. The borrower cannot pass the grant funds directly through to the project, but rather must establish a revolving fund from which to make zero interest rate loans. Loans derived from the initial funding can only be used for specific purposes, but funds generated by the repayment of the initial loans can be used for any rural development project.

Congress appropriated $500,000 for the program in fiscal year 1989. The REA issued its final rules for the program's operation in February 1989 and selected its first loans recipients in May 1989, including two telephone borrowers: The Pine Telephone Company, Wright City, Oklahoma, and Cheyenne River Sioux Tribe Telephone Authority, Eagle Butte, South Dakota.

The legislation creating the program established a Rural Economic Development Subaccount within the Telephone Revolving Fund. In addition to any appropriations and loan repayments, the subaccount is credited with an amount calculated based on the cushion of credit payments made. While the loans are funded through the congressional appropriations, grants are made using the cushion-of-credit-based credits to the subaccount.

With the reorganization of the Department of Agriculture in 1994, operation of the Rural Economic Development Loan and Grant programs shifted to the Rural Business-Cooperative Service. The loans and grants still are made to RUS borrowers, however, the application and approval process has changed and now requires initial submission to the U.S. Department of Agriculture Rural Development State Office.

Federated Telephone Cooperative used another Rural Economic Development Loan to allow a local resident to establish Appletree Graphics, a typesetting business employing 16 individuals on a part-time basis.

FIRE STATION

Panhandle Telephone Cooperative, Guymon, Oklahoma, stands as one of the pioneers in the area of interactive television (ITV) for education. Its Panhandle Shared Video Network, pictured here, began operation in 1988 and served as a model for many such systems that followed. In 1989, Panhandle representatives conducted a presentation on the system and its operation at the REA headquarters for REA staff, as well as staff of the national trade associations. ITV, often called distance learning, promotes rural economic development by allowing small rural schools to share teachers and combine students to justify offering advanced courses. Often the systems also are used for adult education courses.

Improving Operations

Continuing the trend of promoting economic development in rural America, Senator Patrick Leahy (D-VT) introduced a comprehensive rural development bill in spring 1989. As written, the measure included several provisions that would have encouraged bypass and duplication of facilities. But through the work of the National Telephone Cooperative Association and other industry associations, these provisions were dropped, and the bill passed the Senate in August.

Meanwhile, the House rural development package, developed by Representative Glenn English (D-OK) and Representative Tom Coleman (R-MO) and introduced later in the year, took a much different approach. Included in the House package was a telecommunications measure, developed by Representative Bob Wise (D-WV), that not only gave the REA new rural development authority, but also addressed several REA operational matters. The House bill sponsors considered improved use of the existing resources within the REA program to be a key part of rural economic development efforts. The bill passed the House in March 1990.

Ultimately, both the House and Senate attached their respective rural development bills to their versions of the 1990 Farm Bill, which passed Congress in the fall with the majority of the REA provisions intact. President George Bush signed the bill into law on November 28, 1990.

The legislation revised numerous REA policies and practices, and made several modifications to the Rural Telephone Bank (RTB). Among the specific provisions: a permanent prohibition against the REA's general funds policy, an updated definition of telephone service, a requirement that the REA lend for all act purposes and for the full 35-year loan period, and establishment of separate telephone and electric revolving funds.

In terms of the RTB, the measure changed the board structure—increasing from five to seven those named by the President; prohibiting REA staff from holding the government seats on the board; and allowing for the election of the board chair. It also added voting by mail to increase participation in the RTB board elections and changed the required number of votes for a candidate from a majority to a plurality.

As for the new rural development initiatives for the REA, the Farm Bill created three programs, including the Distance Learning and Telemedicine Program, but none of them received funding for 1991. The law also created the position of assistant administrator for economic development within the REA.

Facts and Figures: *REA/RUS Program Statistics*

The following figures are from *1997 Statistical Report: Rural Telecommunications Borrowers*, published in August 1998. The information, which reflects the program's status as of December 31, 1997, was the most recent data available at the time of publication. All dollar amounts are in millions and rounded. (Non-government supplemental lenders also have loaned funds to REA/RUS-eligible borrowers, but those numbers are not reflected in these figures.)

Number of Borrowers

RUS	989
Rural Telephone Bank[1]	662

Financing and Loans Approved—Cumulative Totals[2]

Total Financing	$12,116.2
Total RUS Loans	$7,661.6
2 percent	$2,929.7
5 percent[3]	$3,650.6
Hardship 5 percent	$277.8
Cost of Money	$803.4
Rural Telephone Bank Loans	$3,481.8
RUS Guarantee Commitments	$972.9

Funds Advanced—Cumulative Totals[2]

Total Financing	$9,740.8
Total RUS Loans	$6,430.8
2 percent	$2,928.3
5 percent[3]	$3,222.9
Hardship 5 percent	$99.4
Cost of Money	$180.3
Rural Telephone Bank Loans	$2,626.8
RUS Guarantee Commitments	$683.2

Principal and Interest Payments—Cumulative Totals[4]

RUS Loans	
Principal	$2,965.3
Interest	$2,506.0
Rural Telephone Bank Loans	
Principal	$1,208.5
Interest	$2,055.4
Guaranteed Loans—Federal Financing Bank	
Principal	$330.5
Interest	$680.6

[1] All but 37 RTB borrowers are included in RUS borrowers.

[2] Rescissions are deducted from these amounts.

[3] Includes three variable rate loans.

[4] Includes notes paid in full.

The first few years of the 1990s brought much of the same, with disagreement between the industry and the administration about how the program should operate. But by late 1993, things began to change. The REA took on a new form, a new focus, even a new name. And with the passage of the Telecommunications Act of 1996, the agency's role changed even more. Program funding levels reached new highs, and both the REA and its borrowers entered the world of the Internet and the World Wide Web. As it approached its 50th anniversary, the REA telephone loan program was ...

Evolving and Expanding

By the end of 1991, the REA had provided telephone borrowers more than $1.1 billion for digital switching equipment. One of the digital switches most commonly used in REA borrower exchanges then, and still today, is the Nortel Networks (formerly Northern Telecom) DMS-10.

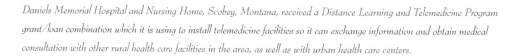

Linking for Learning and Health Care

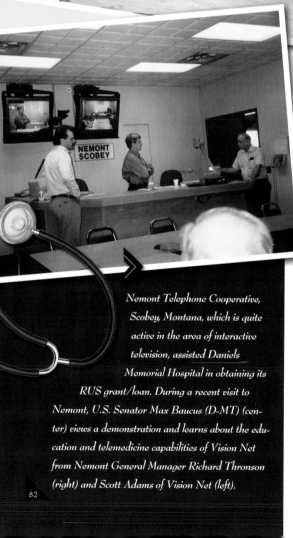

Nemont Telephone Cooperative, Scobey, Montana, which is quite active in the area of interactive television, assisted Daniels Memorial Hospital in obtaining its RUS grant/loan. During a recent visit to Nemont, U.S. Senator Max Baucus (D-MT) (center) views a demonstration and learns about the education and telemedicine capabilities of Vision Net from Nemont General Manager Richard Thronson (right) and Scott Adams of Vision Net (left).

O ne of the rural development programs created by the 1990 Farm Bill—the Distance Learning and Telemedicine (DLT) Program—blossomed in the 1990s, emerging not only as a viable rural development effort, but also as a key component in achieving the national goal of quality education and health care for all rural Americans. The program provides grants—and, as of 1996, loans—to rural schools, libraries, hospitals, and health care providers so they can purchase telecommunications facilities and equipment for distance learning and telemedicine projects.

The DLT program is for equipment and facilities; system operating expenses generally are not eligible. But that's where the program's perfect partnership with the e-rate comes in. Established as part of the universal service provisions of the Telecommunications Act of 1996, the e-rate provides 20 to 90 percent discounts for telecommunications services—including internal connections and recurring monthly service costs—to schools, libraries, and rural health care providers. Together, the e-rate and the DLT program add up to an effective way to enhance rural education and health care.

The REA issued its final rules for the grant portion of the DLT program in late February 1993 and made its first grants in September 1993. The grants do not flow through RUS borrowers, but rather go directly to the education or health care facility. Borrowers are, however, encouraged to take an active role in creating an awareness of the grants in their communities. The 1996 Farm Bill reauthorized the grant portion and established the loan segment of the program, which made its first allocation in October 1997. RUS borrowers can obtain a loan on behalf of a school or medical system.

The program offers three types of financing—grant only, loan only, and a combination of grant and loan. Grant funding is awarded on a competitive basis to all eligible applications received during the filing window each fiscal year, while loans and combination grant/loans are available throughout the year and processed on a first-come, first-served basis. While the program concentrates on equipment and facilities, the RUS will consider items, such as software and training, that are part of the cost of establishing the project.

Through fiscal year 1998, the DLT program had funded 252 projects in 43 states and two U.S. territories, totaling $62.5 million in grants and $3 million in loans. More than 1,000 schools and learning centers and 725 hospitals and rural health clinics have been the recipients of the funds.

During the 1999 National Telephone Cooperative Association (NTCA) Annual Meeting, Executive Vice President Michael Brunner updates members on industry matters, including RUS and Rural Telephone Bank (RTB) issues, and encourages them to talk to their members of Congress. Throughout the various program crises is the early 1990s, NTCA and its members worked with Congress to restore the REA telephone program, and their effective grass-roots lobbying efforts continue to this day.

More Dark Days

A t one point, the events of the early 1990s brought to mind those of late 1972, as unexpected administration actions halted the REA telephone loan program. But this time instead of a shift in the program funding source, the administration's moves brought a total stoppage of loans.

The first shutdown action came in August 1990 when REA Administrator Gary Byrne notified borrowers in Indiana that the REA could not lend any further monies in the state pending resolution of its dispute with the Indiana Utility Regulatory Commission. The matter centered on a repayment problem involving an electric cooperative. Rulings in lawsuits filed in various federal courts, as well as an opinion handed down by the Indiana Supreme Court, proved unfavorable to the REA. Byrne said the agency must place a moratorium on all loans and loan processing until the state could prove the REA's funds in the state were secure.

Although the REA and Indiana regulators reached a tentative compromise later in the year, and the Indiana commission subsequently issued an order exempting the REA from the restriction in question, Indiana borrowers were on hold for several months before the entire situation resolved.

With the Indiana crisis still in full swing, Byrne made another announcement in November 1990 that affected all borrowers. The administrator said that provisions in the 1990 Farm Bill forced him to suspend all REA and RTB loan approvals until the regulations implementing the new law were in place. He explained that the Office of General Counsel for the REA had advised him to take such action. The agency continued to process loans but did not go through the final approval process.

Just days before Byrne's announcement, when signing the Farm Bill into law, President George Bush included in his remarks strong criticisms of the REA portion of the measure. He said the REA modifications "represent unwarranted increases in Federal subsidies and risk" and the provisions allowing loans for all act purposes and for the full 35-year amortization period "in effect turn control of the program over to the borrowers."

While Byrne took extensive heat from Congress for his actions, the national moratorium lasted until mid-June when the agency published its final regulations implementing the Farm Bill.

During the National Telephone Cooperative Association Legislative Conferences, members from North Dakota always take time to visit urban legislators to explain issues such as the importance of RUS funding. Doug Tansey, legislative assistant to Representative Bob Franks (R-NJ) (left), meets with North Dakota members (clockwise from left) David Dunning, manager of Polar Communications, Park River; Rita Greer; Ralph Greer, director of Dickey Rural Telephone Cooperative, Ellendale; and Wayne Lemer, plant manger of West River Telecommunications Cooperative, Hazen, during the 1999 conference.

United States Department of Agriculture
Rural Electrification Administration

February 1991

RETURN TO THE MISSION:
The Administration's Budget and Legislative Proposals for the Future of the Rural Electrification Administration

During his speech at the 1991 National Telephone Cooperative Association Annual Meeting, REA Administrator Gary Byrne discusses the general funds policy, as well as several provisions of the 1990 Farm Bill.

Lots of Negativity, But Little Effect

By early 1991, a full-blown anti-REA campaign had erupted, including a rash of negative media coverage. On February 4, nationally syndicated columnist James Kilpatrick characterized the REA program as one in which "a few rural telephone enterprises are royally milking the taxpayers" and specifically attacked loans to holding companies, as well as funding for buildings and vehicles.

REA Administrator Gary Byrne echoed these same accusations, as well as made comments about the general funds policy and the 35-year amortization period, in his February 1991 speech to the National Telephone Cooperative Association (NTCA). He stated that provisions of the 1990 Farm Bill made the program "look irresponsible" and conflicted with "core mandates of the Rural Electrification Act." He advocated returning to the REA's "original mission"—"funding to build infrastructure in rural America ... not to provide subsidized financing to anyone."

In March, the administration followed with *Return to the Mission: The Administration's Budget and Legislative Proposals for the Future of the Rural Electrification Administration.* The booklet supported the administration's fiscal year 1992 budget proposal to move toward private capital and called for rescinding several Farm Bill provisions which it said "increase the government's credit risk" and move the agency toward areas "in which it has little experience." Congress was quite critical of the publication, not only for its content, but also for its likely being an illegal lobbying tool, as well as a waste of money.

The media attacks picked up again in May when a front-page article in *The Wall Street Journal* charged REA telephone borrowers with being "flush with cash." A month later, a follow-up story discussed an REA list of more than 100 rural telephone systems that had "built up large sums of cash or invested heavily in other ventures." Congress held several hearings on the administration's anti-REA campaign. While it did ask the General Accounting Office to review allegations that borrowers had excessive cash, Congress stood behind the telephone program.

One of the biggest media assaults on the telephone program came on October 27, 1991, when CBS newsmagazine *60 Minutes* broadcast a report on the REA titled "Welfare for the Wealthy." The show aired opposite the seventh game of the World Series in some parts of the country and against a Redskins game in the Washington, D.C., market, so REA critics did not get the exposure they had anticipated.

Unflattering media coverage continued to surface on and off during the next few years, but all with little impact on the program. Much credit for the lack of effect on the program goes to the borrowers, NTCA, and the other industry trade associations. Each time the negativity surfaced, the industry responded with press releases, letters to the editor, or letters to Congress, correcting misrepresentations and pointing out the merits of the telephone loan program.

RECORDING

A Change Supported by All

1993 brought a democrat to the White House for the first time in 12 years, but initially the song appeared to be the same. In his first State of the Union address, President Bill Clinton pointed to the REA: "I recommend that we reduce interest subsidies to the Rural Electrification Administration." He later proposed basically eliminating the 5 percent loan program.

Realizing administration attempts to terminate or severely modify the program were not going to halt, Representative Glenn English (D-OK), chairman of the House agriculture subcommittee with jurisdiction over the REA, decided the time had come to offer a viable alternative for the program, one that would be acceptable to both the industry and the administration. English began drafting a legislative package, seeking input from both sides, and what emerged was the Rural Electrification Loan Restructuring Act of 1993 (RELRA).

Originally part of the budget reconciliation measure, RELRA later was introduced as a separate measure and quickly passed both the House and Senate on voice votes. On November 1, 1993, the President signed the measure into law. The package, which succeeded in winning the support of both REA critics and proponents, implemented a new structure and new lending criteria for the REA loan programs.

Basically, the law established a four-piece REA telephone loan program: REA hardship loans at 5 percent interest; REA cost-of-money loans at the government's cost of money; Rural Telephone Bank (RTB) loans with interest rates calculated based on a formula; and REA loan guarantees. The law eliminated the 2 percent hardship loan program.

The law set forth specific lending criteria for each type of loan, including the allowed uses for loan funds and borrower eligibility requirements. It also included an inflation indexing provision allowing for automatic increases in loan levels based on increases in the Consumer Price Index. The measure additionally raised the population cap on what was considered a rural area from 1,500 to 5,000.

RELRA implemented a new requirement for all REA borrowers—the state telecommunications modernization plan (STMP). To continue to be eligible to borrow from any of the REA programs (except REA loan guarantees), telephone systems had to participate in an REA-approved modernization plan. The law required that each state public utilities commission or legislature develop such a plan. In the alternative, the majority of current REA borrowers in a state could work together to develop the plan.

The RELRA established various service and facility requirements and goals to be included in the STMP, which basically required a design easily upgradeable to broadband services to the home, clearing the way for the network of the future. Of the 51 jurisdictions with REA borrowers, 46 had modernization plans approved as of late 1998.

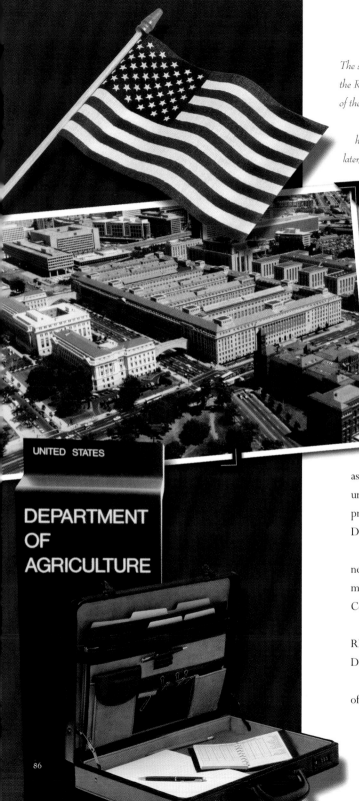

The sprawling U.S. Department of Agriculture building along Independence Avenue has not always been home to the REA—and neither has Washington, D.C. As the program began in 1935, the staff worked from the basement of the U.S. Department of the Interior. In late May of that year, the REA moved to a mansion at 2000 Massachusetts Avenue, the former home of George Westinghouse Jr. As the agency grew, satellite offices sprang up in eight town-houses nearby. In early 1941, the REA made a short move to 1201 Connecticut Avenue. But less than a year later, in December 1941, World War II sent REA employees to the Boatmens National Bank Building in St. Louis, Missouri, to free up Washington office space for those government agencies directly involved in the war. Upon the conclusion of the war, the REA moved back to Washington and into the U.S. Department of Agriculture building, which remains its home to this day. It has, however, weathered some attempts to again relocate it.

A New Name

As part of an overall effort to streamline the federal government, Secretary of Agriculture Mike Espy in September 1993 announced a proposal to reorganize the Department of Agriculture. Just over a year later, Congress passed the Department of Agriculture Reorganization Act of 1994, and the President signed the measure into law. The implementation of the new law brought major changes for the REA, including a new name.

Effective December 1, 1994, the telephone and electric loan programs of the REA, as well as the Rural Telephone Bank and Distance Learning and Telemedicine Program, were placed under the newly created Rural Utilities Service (RUS). The RUS encompasses not only these loan programs of the former REA, but also the water and sewer programs previously under the Rural Development Administration (RDA). The REA administrator became the administrator of the RUS.

The REA's Rural Economic and Development Loan and Grant programs shifted to the new Rural Business-Cooperative Service. Also included in this agency are RDA business development programs, the Agriculture Cooperative Service, and the Alternative Agricultural Commercialization Center.

The REA's rural community loan programs were combined with similar programs from the RDA and with the Farmers Home Administration to form the Rural Housing and Community Development Service.

The undersecretary for rural economic and community development oversees the activities of all three agencies.

The Privatization Process Begins

T he 1990s proved a decade of uncertainty for the Rural Telephone Bank (RTB), as the issues of credit reform and privatization of the bank both presented many unanswered questions. The passage of the Federal Credit Reform Act in 1990 brought a significant change for the structure of the bank. Under this legislation, all federal loan programs were reorganized into a liquidating account and a financing account.

For the RTB, the liquidating account accepts payments on, and makes advances for, loans and commitments made prior to fiscal year 1992. The financing account is the source for all new loans made and accepts payments on those loans. The financing account borrows loan funds from the U.S. Treasury and also accepts a subsidy from Congress to handle administrative expenses and to make up any interest rate differential. The RTB no longer has access to its own low-cost embedded capital, but rather has to borrow from the U.S. Treasury at new higher rates. Under credit reform, any cash remaining in the financing account after the bank meets its obligations at the close of each fiscal year is subject to confiscation by the U.S. Treasury.

For the general REA telephone program, credit reform meant the end of the revolving fund, with the program now having permanent authority to meet its cash requirements from the U.S. Treasury. This proved beneficial in that the budgetary cost of the program became just the subsidy—the differential between the government's cost of money and the interest rate borrowers pay—not total outlays.

As credit reform went into effect on October 1, 1991, the telephone industry trade associations worked with Congress to secure an exemption for the RTB, but due to a budgetary technicality, the attempt was not successful. While the bank did reorganize into the two accounts, the U.S. Treasury has not swept any remaining funds due to appropriations language preventing such transfers. A big unknown is whether credit reform will continue to apply to the RTB once the bank is privatized.

The RTB began privatization as scheduled at the end of fiscal year 1995 and is slowly retiring the government's Class A stock. Appropriations language has limited the amount of Class A stock that can be retired each year.

The exact appearance of the privatized RTB remains to be seen. Some believe the RTB could start lending for purposes from which it currently is restricted under the Rural Electrification Act. Others suggest that once the bank is privatized, it will not be subject to federal requirements and mandates, including annual lending limits and interest rate formulas. But once it is no longer a government entity, the bank also likely will not be able to use government staff and resources without paying for them.

Presidential budget proposals in recent years have called for accelerated privatization, but thus far have not been successful. The most recent proposal calls for the RTB to become a performance-based organization (PBO) as an interim step to privatization. This would allow the bank to operate to a large degree like a private organization.

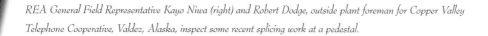

REA General Field Representative Kayo Niwa (right) and Robert Dodge, outside plant foreman for Copper Valley Telephone Cooperative, Valdez, Alaska, inspect some recent splicing work at a pedestal.

Not the Typical Agency

L ike the rural telephone system people with whom they work, the REA/RUS staff through the years have dedicated themselves to bringing the highest quality telecommunications services to the citizens of rural America. From the beginning, the REA structure did not fit the typical government agency mold. The staff includes not only career government employees, but also many individuals from the telephone industry. The agency's engineers arguably are the best in the government, allowing the REA to take a leadership role in the development of rural telephony.

The REA also operates within the unique framework of having field representatives—employees scattered across the nation who work from their homes and travel to the borrowers, providing necessary technical and management assistance. To many borrowers in the early days, these field representatives were the REA.

As the young telephone systems began their operations, agency staff members traveled from system to system, attending organizational meetings, reviewing loan documents, setting up workshops, developing system designs, and managing countless other activities. Many visited each borrower in their area every few weeks. These field agents worked tirelessly, just like the borrowers.

As the years passed, the REA staff members' visits became less frequent and the issues of concern evolved, but the dedication and helpfulness remained the same. Borrowers tell of field representatives assisting their staffs in getting over their fear of computers and helping them understand new regulations and legislation.

During field trials, REA staff were on hand to measure the performance of the equipment, and this often was not an easy job. So as not to disrupt service, many tests happened late at night. This meant not only long hours, but also unique opportunities such as being out in subzero temperatures during the winter. Addressing equipment performance problems brought similar situations, like being left alone at a distant site on the open range for a full day of testing.

Working side by side, the REA staff members and borrowers often become personal friends. Borrowers fondly remember REA staff from the early days, commenting how in many cases they almost were like a member of the family. The friendships develop not only with the field staff, but also with those in Washington, D.C. REA Administrator David Hamil on several occasions referred to his visits and friendships with borrowers.

The staff's work with the borrowers proves rewarding as well. One former loan examiner, who worked in both the telephone and electric programs, commented on the unique working relationship. "The companies sort of looked up to you, and you had a sense of possessiveness about your companies. What was great about the job was seeing the tangible results of your work. You knew you were not just pushing papers."

Today, the number of RUS field representatives and other staff is much fewer, but they are there to assist borrowers just the same.

Northeast Area Field Representative Paul Henderson (left) discusses the preparation of a new requisition with Donald Gruneisen, vice president of Nicholville Telephone Company, Nicholville, New York.

New Exchanges, New Companies

Construction crews for Scott County Telephone Company install trunks connecting the telephone company's facilities to GTE's access tandem in Waldron, Arkansas.

While the REA/RUS's work generally has evolved a long way from addressing the lack of service in the 1950s and the major upgrades of the 1960s and 1970s, the agency continued to fund the formation and upgrading of new exchanges and even new telephone companies in the 1990s.

Many of the additional exchanges, as well as several new telephone companies, resulted from an industry trend completely opposite of the problematic trend of the early 1900s—instead of the Bell companies buying independent telephone systems, independent systems started purchasing rural Bell exchanges.

As they faced increasing competition, the Bell companies, as well as several large independents, determined upgrading their rural exchanges would not be a profitable venture. Thus, they put them up for sale. This presented nearby independents the opportunity to expand their operations and, in some instances, gave telephone systems, and even individuals, the chance to form new telephone companies.

Often the telephone systems came to the REA/RUS to finance the necessary upgrades of the newly acquired territories, the majority of which had extremely poor facilities.

But not all of the additions came from large company exchange sales. In August 1991, for instance, the REA provided funding to a newly formed company, the Scott County Telephone Company, to provide first-time telephone service to more than 150 families in northwestern Arkansas. The 250-square-mile area, a large portion of which fell within the Ouachita National Forest, had no service due to the cost of installing and maintaining service in such a remote location with mountainous terrain. Scott County used its $1.5 million REA loan to install a digital central office and 100 miles of buried plant. The company initiated service in 1993.

At the official groundbreaking for Scott County Telephone Company are (front row, from left) Tommy Dorries, owner; Representative W.R. "Bud" Rice of Waldron; Robert Prince, owner; (back row, from left) Bryan Howerton, REA; Butch Nelson, Klaasmeyer Construction Company; Larry Redding, C.H. Guernsey Engineering; Ron Newton, C.H. Guernsey; Jackie Tedford, Arkansas Public Service Commission; and Stan Joyner, Klaasmeyer. (Not pictured: Bobby Dorries, owner.)

Engineering department employees at Penasco Valley Telephone Cooperative, Artesia, New Mexico, use the latest computer software for mapping and designing the cooperative's outside plant facilities. Pictured are (from left) Laura Weddige, engineering technician; Gary Nelson, assistant engineer; and Yolanda Agnew, engineering technician.

The State of Technology

The emerging technologies and services of the 1990s alone had the virtual alphabet soup of telecommunications industry acronyms overflowing. From ADSL and HDSL to SONET and FTTH,* the possibilities continued to expand. The RUS worked with manufacturers and conducted borrower field trials and also kept in close contact with the various standards-setting organizations.

As the Internet and World Wide Web came on the scene, bringing many new challenges and opportunities for RUS borrowers, the agency itself took advantage of this new communications medium. In April 1996, the RUS launched its web site and in mid-1997, added its "List of Materials Acceptable for Use on Telecommunications Systems of RUS Borrowers." With the list posted on the web site, people no longer have to subscribe, and the most up-to-date list is available within days of the monthly approval committee meeting. The RUS also launched a special children's web site in May 1998, offering a fun, yet educational, look at the agency and the utility industries it serves.

When a mid-1990s study revealed that the National Oceanic and Atmospheric Administration (NOAA) emergency weather alerting system only covered about 75 percent of the nation's land area, the RUS launched a new project. The agency joined the NOAA and the Federal Emergency Management Agency in developing a plan to increase and improve the system's coverage. The RUS identified borrowers in limited coverage areas and began contacting them requesting that they consider donating tower space, building space, communications facilities, electrical power, or funds to purchase a transmitter for their service areas. The project has been quite successful.

Also, with the pending arrival of the year 2000 and the associated concerns about its potential impact on computer and communications systems, the RUS surveyed and worked with its borrowers to help ensure their readiness.

Throughout the 1990s, the RUS undertook a major effort of streamlining regulations and agency operations to eliminate much of the "back and forth" between the agency and borrowers, shorten loan processing times, and give borrowers more responsibility and flexibility. And by the mid-1990s, administration support for the program had grown strong, with budget proposals being higher than they had been in many years.

*asymmetrical digital subscriber line, high-bit-rate digital subscriber line, synchronous optical network, fiber to the home

Not long after being named REA administrator, Wally Beyer speaks at the 1994 National Telephone Cooperative Association Annual Meeting. Since that time, Beyer's title has changed to RUS administrator and the agency he oversees has significantly changed as well. Recently Beyer outlined the current RUS focus on increasingly becoming a provider of credit support, leveraging private investment via guarantees instead of being the lender per se. The agency already has developed guarantee arrangements with the private lenders serving the rural telecommunications industry.

A Whole New World

1 996 brought one of the most significant events in telecommunications history—the passage of the Telecommunications Act of 1996. This measure ushered in a new era for rural telecommunications and new opportunities for rural Americans. Like divestiture, the Telecom Act had a much greater impact on small, rural telephone systems than many believed it would have. And with the law and its many implementation issues came some changes for the RUS.

Starting in the mid-1990s, the agency had begun to view itself as the administration's spokesperson for rural infrastructure development, representing the interests of all of rural America, not just the viewpoint of its borrowers. This new advocacy role aimed at ensuring quality telecommunications services and the necessary infrastructure are available for all rural Americans, even those its borrowers do not serve.

With the passage of the Telecom Act, the RUS found itself significantly stepping up this advocacy role. The act meant increased interaction with the Federal Communications Commission (FCC), not only in terms of filings on act implementation issues, but also meetings with the regulators. Because the act gives the states authority in many matters, the RUS also expanded its relationships with state regulators, both individually and through their association, the National Association of Regulatory Utility Commissioners.

The RUS's interaction with other government agencies increased as well. RUS representatives now regularly participate in several interagency meetings on telecom matters including weekly Telecommunications Working Group and E-Rate Working Group meetings, both at the White House. Also, an RUS representative, as well as several borrowers, serve on the FCC's Rural Task Force, which advises both the FCC and the Federal-State Joint Board on universal service issues and helps these regulators determine how decisions can be adapted for rural telephone systems.

Because the Telecom Act opens local telecommunications markets to competition, it presents some challenges to the RUS in terms of reconciling the act's provisions with those of the Rural Electrification Act. But in general, the RUS leadership views the Telecom Act as a plus, highlighting the language that assures rural Americans will be included in the information age, having access to advanced services at reasonable rates.

In a commentary in the *National Telephone Cooperative Association's* Phone Call *magazine* concerning the 25th anniversary of the REA telephone loan program, then-NTCA President Harold Ebaugh looked ahead. "On the 50th anniversary in 1999 (while I don't imagine many of us will still be involved in the program), I'm confident that the program will be going forward." While he was a bit wrong on the involvement issue, he definitely was right about the program.

As the century drew to a close, RUS telecommunications borrowers had used more than $12 billion in approved financing to build a rural network of more than one million miles of line carrying service to millions of rural residents. But the agency's mission is far from complete.

The program began due to the need for capital, and that need still exists, albeit for different purposes. Today, simply having a telephone is not enough. Every American deserves access to advanced telecommunications services, and that access should be high-quality, reliable, and at a reasonable price. The RUS remains committed to providing the capital and assistance rural telephone systems need to operate and expand in the ever-evolving telecommunications industry.

The RUS telecommunications program operates at little cost to the government—the annual subsidy cost is a fraction of the total capital the program provides. And the benefits to the nation go far beyond the communications services borrowers bring, with increased excise taxes and income taxes, as well as many employment and rural economic development opportunities.

But perhaps the greatest testament to the success of the program and its borrowers' commitment to rural service is the telephone loan program's perfect record—in the agency's 50-year history, there never has been a default on a telephone loan.

As the new century approaches, the telecommunications industry is in the midst of an exciting time. The RUS and its borrowers must learn from the past and focus on the future as they move …

Into the Next Millennium

The Foundation for Rural Service (FRS) wishes to thank the following telephone companies and cooperatives who provided their stories, documents, and photographs in response to our requests for information. Although we were not able to include every item in the final publication, all contributions were greatly appreciated.

The Arthur Mutual Telephone Company, Defiance, Ohio

Bristol Bay Telephone Cooperative, King Salmon, Alaska

Bruce Telephone Company, Bruce, Mississippi

Cameron Telephone Company, Sulphur, Louisiana

Chequamegon Telephone Cooperative, Cable, Wisconsin

Cheyenne River Sioux Tribe Telephone Authority, Eagle Butte, South Dakota

Clarks Telecommunications Company, Clarks, Nebraska

Commonwealth Telephone Company, Dallas, Pennsylvania

Consolidated Telephone Cooperative, Dickinson, North Dakota

Dunnell Telephone Company, Dunnell, Minnesota

Egyptian Telephone Cooperative Association, Steeleville, Illinois

Emery Telephone, Orangeville, Utah

Farmers Mutual Telephone Cooperative, Shellsburg, Iowa

Federated Telephone Cooperative, Chokio, Minnesota

Fort Mojave Telecommunications, Mohave Valley, Arizona

Garden Valley Telephone Company, Erskine, Minnesota

Gila River Telecommunications, Chandler, Arizona

Glenwood Telephone Membership Corporation, Blue Hill, Nebraska

GT Com (formerly Florala Telephone Company), Florala, Alabama

Hardy Telecommunications, Lost River, West Virginia

Lafourche Telephone Company, Larose, Louisiana

Leaco Rural Telephone Cooperative, Lovington, New Mexico

Loretto Telephone Company, Loretto, Tennessee

Midstate Telephone Company, Kimball, South Dakota

Mutual Telephone Company, Sioux Center, Iowa

Nelson Telephone Cooperative, Durand, Wisconsin

Nemont Telephone Cooperative, Scobey, Montana

Nushagak Telephone Cooperative, Dillingham, Alaska

OTZ Telephone Cooperative, Kotzebue, Alaska

Panhandle Telephone Cooperative, Guymon, Oklahoma

Park Region Mutual Telephone Company, Underwood, Minnesota

Penasco Valley Telephone Cooperative, Artesia, New Mexico

Piedmont Rural Telephone Cooperative, Laurens, South Carolina

Plant Telephone Company, Tifton, Georgia

Randolph Telephone Company, Liberty, North Carolina

Reservation Telephone Cooperative, Parshall, North Dakota

Ringgold Telephone Company, Ringgold, Georgia

Roosevelt County Rural Telephone Cooperative, Portales, New Mexico

S&A Telephone Company, Allen, Kansas

San Carlos Apache Telecommunications Utility, San Carlos, Arizona

Scott County Telephone Company, Valliant, Oklahoma

South Central Telephone Association, Medicine Lodge, Kansas

TelAlaska, Anchorage, Alaska

Tohono O'Odham Utility Authority, Sells, Arizona

The Tri-County Telephone Association, Council Grove, Kansas

Waitsfield-Fayston Telephone Company, Waitsfield, Vermont

Winnebago Cooperative Telephone Association, Lake Mills, Iowa

XIT Rural Telephone Cooperative, Dalhart, Texas

The following organizations and individuals also provided documents and photographs:

8x8 Inc.

Claude Buster

Patrick Dahl

Daniels Memorial Hospital

Margaret Goatcher

Lamar Moore

The Museum of Independent Telephony

National Rural Electric Cooperative Association (NRECA)

National Rural Telecom Association (NRTA)

National Telephone Cooperative Association (NTCA)

Nortel Networks

Organization for the Promotion and Advancement of Small Telecommunications Companies (OPASTCO)

Parker Telecommunications

Rural Telephone Finance Cooperative

Rural Utilities Service

Wisconsin State Telephone Association

Photograph Credits

E*very effort has been made to properly identify and give credit to the photographers of the images contained in this publication. Images credited to "USDA Photo" are U.S. Department of Agriculture (USDA) or Rural Electrification Administration (REA) publicity photographs marked with that specific credit or are from the USDA or REA records at the National Archives. The majority of images without credits came from the Foundation for Rural Service (FRS) and National Telephone Cooperative Association (NTCA) photograph files, or were supplied by one of the project participants (see list on page 93) without a credit.*

Page 11: .... USDA Photo

Page 13: .... (top) USDA Photo; (bottom) A.L. Gaver, USDA Photo

Page 14: .... (both) USDA Photo

Page 15: .... USDA Photo

Page 16: .... (top) J.T. Holloway; (both bottom) USDA Photo

Page 17: ....(both) USDA Photo

Page 18: ....(top) USDA Photo; (map) USDA Photo

Page 19: ....USDA Photo

Page 20: .... Worldwide Photos

Page 22: .... USDA Photo

Page 23: .... (top) A.L. Gaver, USDA Photo; (bottom) *Brule County News*, Kimball, South Dakota, May 26, 1955

Page 24: .... (top) Global Productions, Las Vegas, Nevada; (bottom) USDA Photo

Page 25: .... (bottom) A.L. Gaver, USDA Photo

Page 26: .... (poster) USDA Photo

Page 27: .... (top) USDA Photo; (bottom) *Farmers Educational and Cooperative Union*, Denver, Colorado

Page 28: .... USDA Photo

Page 30: .... A.L. Gaver, USDA Photo

Page 31: .... (top) USDA Photo

Page 32: .... (both) USDA Photo

Page 33: .... (bottom) USDA Photo

Page 34: .... (both) USDA Photo

Page 37: .... Capitol and Glogau, Washington, DC

Page 38: .... (both) USDA Photo

Page 39: .... (top) USDA Photo

Page 40: .... (top) A.L. Gaver, USDA Photo; (bottom) USDA Photo

Page 41: .... USDA Photo

Page 43: ....USDA Photo

Page 44: .... USDA Photo

Page 45: .... (bottom) Christen Studio, Lemmon, South Dakota

Page 47: .... (top) USDA Photo

Page 49: .... USDA Photo

Page 50: .... (both) USDA Photo

Page 52-53:... (all except Ballard and Miller) USDA Photo

Page 55: .... The White House

Page 57: .... (top) USDA Photo

Page 62: .... (top) Jim Felter

Page 68: .... USDA Photo

Page 69: .... (top) USDA Photo

Page 73: .... (top) Gisele M. Byrd

Page 74: .... (top) Fay Photo Service Inc., Boston, Massachusetts

Page 83: .... (bottom) Ron Ceasar

Page 86: .... USDA Photo

Page 89: .... (both) Cindy Lawrence, *Waldron County News*

Page 91: .... Squire Haskins Photography, Dallas, Texas

BIBLIOGRAPHY

I n addition to the extensive information and materials provided by the REA/RUS borrowers and past and present members of the REA/RUS staff, the following publications served as resources for the development of this book.

Publications produced by the Rural Electrification Administration/Rural Utilities Service, U.S. Department of Agriculture:

- *20 Years of Progress in Rural Telephone Service: Breaking the Communications Barrier Between Rural America and the World* (October 1969)

- *1997 Statistical Report: Rural Telecommunications Borrowers*

- *A Brief History of the Rural Electric and Telephone Programs* (FY1990-FY1992)

- *Developing Community Resources* (September 1970)

- *Report of the Administrator* (FY1950-FY1992)

- *Rural America Growth Country* (August 1968)

- *Rural Community Development ... Goals and Actions* (October 1971)

- *Rural Electrification News* (1950-1954)

- *Rural Lines* (1954-1960)

- Rural Telephone Bank Annual Reports (1972-1997)

- *Rural Telephone Service USA: A Pictorial History of Rural Electrification Administration's Telephone Loan Program* (May 1960)

- *Serving Telephone Subscribers in the Year 2000* (June 1977)

- *A Telephone for Your Farm: Answers to Questions About the Rural Telephone Program* (June 1954)

- *Twenty-Five Years of Progress ... Rural Telephone Service USA* (September 1974)

- Also, the Rural Utilities Service web site, the Rural Electrification Act of 1936 as amended, the script of *The Telephone and the Farmer*, and miscellaneous press releases, memos, speeches, and fact sheets.

Publications produced by the National Telephone Cooperative Association:

- *From Squaw Gap to 2626 Pennsylvania Avenue* (1989)

- NTCA Annual Reports (1970-1997)

- *Our First 25 Years - An Informal History* (1979)

- *Phone Call* (1956-1981)

- *Rural Telecommunications* (1982-1999)

- *Washington Report* (misc. issues)

Additional publications:

- *Farmers Telephone Companies: Organization, Financing and Management,* I.M. Spasoff and H.S. Beardsley, U.S. Department of Agriculture, Farmers' Bulletin No. 1245. (1932)

- *A Giant Step,* Clyde T. Ellis (1966)

- *The Next Greatest Thing,* National Rural Electric Cooperative Association (1984)

- *Rural Telephones: Hearings Before the Committee on Agriculture and Forestry, United States Senate, Eighty-First Congress* (June 11 and August 3-5, 8-10, 1949)

- *Rural Telephones: Hearings Before a Subcommittee of the Committee on Agriculture, House of Representatives, Eighty-First Congress* (February 14-16, 1949)